T0231983

Anticipating Correlations

Anticipating Correlations

A New Paradigm for Risk Management

Robert Engle

Princeton University Press
Princeton and Oxford

Published by Princeton University Press,
41 William Street, Princeton, New Jersey 08540

In the United Kingdom: Princeton University Press,
6 Oxford Street, Woodstock, Oxfordshire OX20 1TW

ISBN: 978-0-691-11641-9 (alk. paper)

Library of Congress Control Number: 2008939934

British Library Cataloging-in-Publication Data is available

This book has been composed in LucidaBright using TEX
Typeset and copyedited by T&T Productions Ltd, London

Printed on acid-free paper ∞

press.princeton.edu

Printed in the United States of America

10 9 8 7 6 5 4 3 2 1

Contents

Introduction

The Econometric Institute Lecture Series deals with topics in econometrics that have important policy implications. The lectures cover a wide range of topics and are not confined to any one area or subdiscipline. Leading international scientists in the fields of econometrics in which applications play a major role are invited to give three-day lectures on a topic to which they have contributed significantly.

The topic of Robert Engle's lectures deals with the dynamics of correlations between a large number of financial assets. In the present global financial world it is imperative both for asset management and for risk analysis to gain a better understanding of the changing correlations between a large number of assets and even between different financial markets. In Robert Engle's book several innovative models are proposed and tested with respect to their forecasting performance in the turbulent economic world of 2007.

As editors of the series we are indebted to the Erasmus University Trust Fund, the Erasmus Research Institute of Management, and the Tinbergen Institute for continued support for the series.

Philip Hans Franses and Herman K. van Dijk
Econometric Institute
Erasmus School of Economics

Anticipating Correlations

1

Correlation Economics

1.1 Introduction

Today there are almost three thousand stocks listed on the New York Stock Exchange. NASDAQ lists another three thousand. There is yet another collection of stocks that are unlisted and traded on the Bulletin Board or Pink Sheets. These U.S.-traded stocks are joined by thousands of companies listed on foreign stock exchanges to make up a universe of publicly traded equities. Added to these are the enormous number of government and corporate and municipal bonds that are traded in the United States and around the world, as well as many short-term securities. Investors are now exploring a growing number of alternative asset classes each with its own large set of individual securities. On top of these underlying assets is a web of derivative contracts. It is truly a vast financial arena. A portfolio manager faces a staggering task in selecting investments.

The prices of all of these assets are constantly changing in response to news and in anticipation of future performance. Every day many stocks rise in value and many decline. The movements in price are, however, not independent. If they were independent, then it would be possible to form a portfolio with negligible volatility. Clearly this is not the case. The correlation structure across assets is a key feature of the portfolio choice problem because it is instrumental in determining the risk. Recognizing that the economy is an interconnected set of economic agents, sometimes considered a general equilibrium system, it is hardly surprising that movements in asset prices are correlated. Estimating the correlation structure of thousands of assets and using this to select superior portfolios is a Herculean task. It is especially difficult when it is recognized that these correlations vary over time, so that a forward-looking correlation estimator is needed. This problem is the focus of this book. We must "anticipate correlations" if we want to have optimal risk management, portfolio selection, and hedging.

Such forward-looking correlations are very important in risk management because the risk of a portfolio depends not on what the correlations were in the past, but on what they will be in the future. Similarly, portfolio choice depends on forecasts of asset dependence structure. Many aspects of financial planning involve hedging one asset with a collection of others. The optimal hedge will also depend upon the correlations and volatilities to be expected over the future holding period. An even more complex problem arises when it is recognized that the correlations can be forecast many periods into the future. Consequently, there are predictable changes in the risk–return trade-off that can be incorporated into optimal portfolios.

Derivatives such as options are now routinely traded not only on individual securities, but also on baskets and indices. Such derivative prices are related to the derivative prices of the component assets, but the relation depends on the correlations expected to prevail over the life of the derivative. A market for correlation swaps has recently developed that allows traders to take a position in the average correlation over a time interval. Structured products form a very large class of derivatives that are sensitive to correlations. An important example of a structured product is the collateralized debt obligation (CDO), which in its simplest form is a portfolio of corporate bonds that is sold to investors in tranches that have different risk characteristics. In this way credit risks can be bought and sold to achieve specific risk–return targets. There are many types of CDOs backed by loans, mortgages, subprime mortgages, credit default swaps, tranches of CDOs themselves, and many other assets. In these securities, the correlations between defaults are the key determinants of valuations. Because of the complexity of these structures and the difficulty in forecasting correlations and default correlations, it has been difficult to assess the risks of the tranches that are supposed to be low risk. Some of the "credit crunch" of 2007–8 can probably be attributed to this failure in risk management. This episode serves to reinforce the importance of anticipating correlations.

This book will introduce and carefully explain a collection of new methods for estimating and forecasting correlations for large systems of assets. The book initially discusses the economics of correlations. Then it turns to the measurement of comovement and dependence by correlations and alternative measures. A look at existing models for estimating correlations—such as historical correlation, exponential smoothing, and multivariate GARCH—leads to the introduction (in chapter 3) of the central method explored in the book: dynamic conditional correlation. Monte Carlo and empirical analyses of this model document its performance. Successive chapters deal with extensions to the basic model, new

Table 1.1. Correlations of large-cap stocks from 1998 to 2003.

	IBM	MCD	GE	Citibank	AXP	WMT	SP500
IBM	1.000	0.192	0.436	0.419	0.387	0.283	0.600
MCD	0.192	1.000	0.308	0.238	0.282	0.303	0.365
GE	0.436	0.308	1.000	0.595	0.614	0.484	0.760
Citibank	0.419	0.238	0.595	1.000	0.697	0.439	0.740
AXP	0.387	0.282	0.614	0.697	1.000	0.445	0.715
WMT	0.283	0.303	0.484	0.439	0.445	1.000	0.584
SP500	0.600	0.365	0.760	0.740	0.715	0.584	1.000

estimation methods, and a technical discussion of some econometric issues. Many empirical studies are documented in particular chapters, including stock–bond correlations, global equity correlations, and U.S. large-cap stock correlations. Finally, in a chapter called "Anticipating Correlations," these methods are used to forecast correlations through the turbulent environment of the summer and autumn of 2007.

The methods introduced in this book are simple, powerful, and will be shown to be highly stable over time. They offer investors and money managers up-to-date measures of volatilities and correlations that can be used to assess risk and optimize investment decisions even in the complex and high-dimensional world we inhabit.

1.2 How Big Are Correlations?

Correlations must all lie between -1 and 1, but the actual size varies dramatically across assets and over time. For example, using daily data for the six-year period from 1998 through 2003 and the textbook formula

$$\hat{\rho}_{x,y} = \frac{\sum_{t=1}^{T}(x_t - \bar{x})(y_t - \bar{y})}{\sqrt{\sum_{t=1}^{T}(x_t - \bar{x})\sum_{t=1}^{T}(y_t - \bar{y})}}, \tag{1.1}$$

it is interesting to calculate a variety of correlations. The correlation between daily returns on IBM stock and the S&P 500 measure of the broad U.S. market is 0.6. This means that the regression of IBM returns on a constant and S&P returns will have an R^2 value of 0.36. The systematic risk of IBM is 36% of the total variance and the idiosyncratic risk is 64%.

Looking across five large-capitalization stocks, the correlations with the S&P 500 for the six-year period range from 0.36 for McDonald's to 0.76 for General Electric (GE). These stocks are naturally correlated with each other as well, although the correlations are typically smaller (see table 1.1).

Table 1.2. Correlations of small-cap stocks from 1998 to 2003.

	PVA	NSC	ARG	DRTK	MTLG	SP
PVA	1.000	0.159	0.050	0.063	0.014	0.185
NSC	0.159	1.000	0.253	0.006	0.034	0.445
ARG	0.050	0.253	1.000	0.068	0.081	0.326
DRTK	0.063	0.006	0.068	1.000	0.025	0.101
MTLG	0.014	0.034	0.081	0.025	1.000	0.080
SP	0.185	0.445	0.326	0.101	0.080	1.000

A more careful examination of the correlations shows that the highest correlations are between stocks in the same industry. American Express (AXP) and Citibank have a correlation of almost 0.7 and GE has a correlation with both that is about 0.6. During this period GE had a big financial services business and therefore moved closely with banking stocks.

Examining a selection of small-cap stocks, the story is rather different. The correlations with the market factor are much lower and the correlations between stocks are lower; table 1.2 gives the results. The largest correlation with the market is 0.45 but most of the entries in the table are below 0.1.

Turning to other asset classes let us now examine the correlation between the returns on holding bonds and the returns on holding foreign currencies (see table 1.3). Notice first the low correlations between bond returns and the S&P 500 and between currency returns and the S&P 500. These asset classes are not highly correlated with each other on average.

Within asset classes, the correlations are higher. In fact the correlation between the five- and twenty-year bond returns is 0.875, which is the highest we have yet seen. The short rate has correlations of 0.3 and 0.2, respectively, with these two long rates. Within currencies, the highest correlation is 45% between the Canadian dollar and the Australian dollar, both relative to the U.S. dollar. The rest range from 15% to 25%.

When calculating correlations across countries, it is important to recognize the differences in trading times. When markets are not open at the same times, daily returns calculated from closing data can be influenced by news that appears to be on one day in one market but on the next day in the other. For example, news during U.S. business hours will influence measured Japanese equity prices only on the next day. The effect of the news that occurs when a market is closed will be seen primarily in the opening price and therefore is attributed to the following daily return. To mitigate this problem, it is common to use data that is more time aggregated to measure such correlations.

Table 1.3. Other assets.

	T3M	T5YR	T20YR	CAD/USD	GBP/USD	AUD/USD	JPY/USD	SP500
T3M	1.000	0.329	0.206	0.011	0.076	0.025	0.031	−0.031
T5YR	0.329	1.000	0.875	−0.0007	0.136	0.007	0.005	−0.057
T20YR	0.206	0.875	1.000	0.007	0.103	−0.002	−0.049	−0.016
CAD/USD	0.011	−0.0007	0.007	1.000	0.117	0.415	0.145	0.015
GBP/USD	0.076	0.136	0.103	0.117	1.000	0.253	0.224	−0.018
AUD/USD	0.025	0.007	−0.002	0.415	0.253	1.000	0.269	0.040
JPY/USD	0.031	0.005	−0.049	0.145	0.224	0.269	1.000	−0.003
SP500	−0.031	−0.057	−0.016	0.015	−0.018	0.040	−0.003	1.000

Notes: "T3M" denotes three-month Treasury Bill returns; "T5YR" denotes five-year Treasury bond returns; "T20YR" denotes twenty-year Treasury bond returns; "CAD/USD" Canadian dollar/U.S. dollar returns; "GBP/USD" U.K. pound/U.S. dollar returns; "AUD/USD" Australian dollar/U.S. dollar returns; "JPY/USD" Japanese yen/U.S. dollar returns; "SP500" denotes Standard & Poor's 500 index of equity returns.

Cappiello et al. (2007) analyze weekly global equity and bond correlations. The data employed in their paper consist of FTSE All-World indices for twenty-one countries and DataStream-constructed five-year average maturity bond indices for thirteen, all measured relative to U.S. dollars. The sample is fifteen years of weekly price observations, for a total of 785 observations from January 8, 1987, until February 7, 2002. Table 1.4 shows a sample of global equity and bond correlations. The bond correlations are above the diagonal and the equity correlations are below the diagonal.

The equity correlations range from 0.23 to 0.73 with about a third of the sample above 0.5. The highest are between closely connected economies such as Germany, France, and Switzerland, and the United States and Canada. The bond return correlations are often much higher. France and Germany have a correlation of 0.93 and most of the European correlations are above 0.6. The U.S. correlation with Canada is 0.45, while the correlations with other countries hover around 0.2. Japanese correlations are also lower. Cappiello et al. also report correlations between equities and bonds that vary greatly. Many of these are negative. Typically, however, the domestic equity- and bond-return correlations are fairly large. This is partly due to the fact that both returns are denominated in U.S. dollars.

1.3   The Economics of Correlations

To understand the relative magnitude of all these correlations and ultimately why they change, it is important to look at the economics behind movements in asset prices. Since assets are held by investors in anticipation of payments to be made in the future, the value of an asset is intrinsically linked to forecasts of the future prospects of the project or firm. Changes in asset prices reflect changing forecasts of future payments. The information that makes us change these forecasts we often simply call "news." This has been the basic model for changing asset prices since it was formalized by Samuelson (1965). Thus both the volatilities of asset returns and the correlations between asset returns depend on information that is used to update these distributions.

Every piece of news affects all asset prices to a greater or lesser extent. The effects are greater on some equity prices than on others because their lines of business are different. Hence the correlations in their returns due to this news event will depend upon their business. Naturally, if a firm changes its line of business, its correlations with other firms are likely to change. This is one of the most important reasons why correlations change over time.

Table 1.4. Global equity and bond correlations.

	Canada	Denmark	France	Germany	Ireland	Japan	Sweden	Switzerland	U.K.	U.S.
Canada	—	0.094	0.097	0.068	0.134	0.007	0.160	−0.019	0.167	0.452
Denmark	0.279	—	0.907	0.909	0.839	0.418	0.696	0.800	0.650	0.195
France	0.462	0.496	—	0.927	0.832	0.428	0.679	0.823	0.673	0.267
Germany	0.399	0.399	0.729	—	0.826	0.464	0.657	0.866	0.656	0.221
Ireland	0.361	0.449	0.486	0.515	—	0.359	0.664	0.710	0.699	0.212
Japan	0.230	0.288	0.340	0.321	0.279	—	0.294	0.475	0.343	0.038
Sweden	0.497	0.478	0.577	0.639	0.474	0.317	—	0.553	0.566	0.173
Switzerland	0.391	0.521	0.641	0.722	0.528	0.343	0.549	—	0.589	0.126
U.K.	0.463	0.460	0.594	0.562	0.634	0.350	0.539	0.585	—	0.249
U.S.	0.692	0.299	0.465	0.432	0.392	0.223	0.490	0.398	0.495	—

Notes: equity correlations appear above the diagonal and bond correlations appear below; the figures are for the period 1987–2002.

A second important reason is that the characteristics of the news change. News that has the same qualitative effect on two companies will generally increase their correlation. The magnitude of this news event will determine whether this is an important change in correlations. Consequently, correlations often change dramatically when some factor becomes very important having previously been dormant. An example of this might be energy prices. For years, these fluctuated very little. However, in 2004 prices more than doubled and suddenly many firms and countries whose profitability depended on energy prices showed fluctuations in returns that were more correlated than before (some of these are naturally negative). Thus when the news changes in magnitude it is natural that correlations will change.

Since asset prices of firms are based on the forecasts of earnings or dividends and of expected returns, the movements in prices are based on the updates to these forecasts, which we call firm news. For each asset, we will focus on two types of news: news on future dividends or earnings, and news on future expected returns. Both types of news will depend upon news about energy prices, wage rates, monetary policy, and so forth. Correlations are then based on the similarities between the news for different firms. In particular, it will be shown below that it is correlation between the firm news processes that drives correlation between returns.

To apply this idea to the correlations described in tables 1.1–1.4, it is necessary to show how the underlying firm news processes are correlated. Stocks in the same industry will have highly correlated dividend news and will therefore be more highly correlated than stocks in different industries. Small-cap stocks will often move dramatically with earnings news and this news may have important idiosyncratic components. Consequently, these stocks are naturally less correlated than large-cap stocks. Large-cap stocks will have rather predictable dividend streams, which may respond directly to macroeconomic news. These companies often have well-diversified business models. Hence, volatilities of large-cap stocks should be less than those for small-cap stocks and correlations should be higher. Index returns will also respond to macroeconomic news and hence are typically more correlated with large-cap stocks than with small-cap ones.

For equities, news about the expected return is essentially news about the relevant interest rate for this asset. It will be determined largely by shifts in macroeconomic policy, which determine short rates, and by the risk premium, which in turn will be influenced by market volatility. These effects are presumably highly correlated across stocks within the domestic market. There may be fluctuations across sectors and companies as the actual risk premium could vary with news, but one would expect that this factor would be quite correlated.

The net effect of these two news sources for equities will be a return correlation constructed from each of the basic correlations. The bigger the size of a news event, the more important its influence on correlations will be. Thus when future Federal Reserve policy is uncertain, every bit of news will move prices and the correlations will rise to look more like the correlation in required returns. When the macroeconomy is stable and interest rates have low volatility, the correlation of earnings news is most important. For government bonds there is little or no uncertainty about dividends, but news about the future short-term interest rate is a key determinant of returns. Bonds of all maturities will respond to news on monetary policy or short-term interest rate changes. When this is the major news source, the correlations will be quite high. When there are changes in risk premiums, it will again affect all fixed-income securities, leading to higher correlations. However, when the premium is a credit risk premium, the effect will be different for defaultable securities such as corporate bonds or bonds with particularly high yields. In this case, correlations might fall or even go negative between high-risk and low-risk bonds. Because equities as well as bonds are sensitive to the expected-return component of news, they will be positively correlated when this has high variance. When it has low variance, we might expect to see lower or negative correlations between stocks and treasuries, particularly if good news on the macroeconomy becomes bad news on interest rates because of countercyclical monetary policy.

Exchange rates respond to both domestic and foreign news. If all exchange rates are measured relative to the dollar, there is a natural common component in the correlations. Similarly, international equity and bond returns may be measured in dollar terms. This will increase the measured correlations. Countries with similar economies will have correlated news processes because the same events will affect them. Index returns such as those exhibited in table 1.4 will show more highly correlated returns as the idiosyncratic shocks will be averaged away. Bond returns across countries will generally be highly correlated as the market is truly global; the currency of denomination may be important in flexible exchange rate systems.

1.4 An Economic Model of Correlations

Many of these results are complex and interrelated. Because they have been described in words, the overall simplicity of the argument may not be apparent. In order to put these results in a quantitative context, a mathematical derivation of the correlation of returns needs to be developed. The goal is to show how correlations in returns are based on correlations in the news. We first express continuously compounded returns, r, in terms of the price per share, P, and the dividend per share, D:

$$r_{t+1} = \log(P_{t+1} + D_{t+1}) - \log(P_t). \tag{1.2}$$

Applying the Campbell and Shiller (1988a,b) or Campbell (1991) log-linearization, this can be approximately written as

$$r_{t+1} \approx k + \rho p_{t+1} + (1 - \rho)d_{t+1} - p_t, \tag{1.3}$$

where lowercase letters refer to logs and k is a constant of linearization. Essentially this is a type of series expansion of the log of a sum, which is approximately the weighted average of the logs of the components. The approximation is good if the ratio of the two components is small and relatively constant. These conditions are typically satisfied for equity prices. The parameter ρ is essentially the discount rate and is very slightly below 1.

Solving this equation for p_t, assuming that stock prices do not diverge to infinity, gives

$$p_t = \frac{k}{1 - \rho} + (1 - \rho) \sum_{j=0}^{\infty} \rho^j d_{t+1+j} - \sum_{j=0}^{\infty} \rho^j r_{t+1+j}. \tag{1.4}$$

Taking expectations of both sides with respect to the information at time t gives the same dependent variable, since p_t is in the information set:

$$p_t = \frac{k}{1 - \rho} + (1 - \rho) \sum_{j=0}^{\infty} \rho^j E_t(d_{t+1+j}) - \sum_{j=0}^{\infty} \rho^j E_t(r_{t+1+j}). \tag{1.5}$$

Similarly, taking expectations with respect to information at time $t - 1$ gives the one-step-ahead predictor of prices. The difference between the log of the price expected today and that expected at $t - 1$ is simply the surprise in returns. Hence,

$$r_t - E_{t-1}(r_t) = E_t(p_t) - E_{t-1}(p_t) \tag{1.6}$$

and

$$r_t - E_{t-1}(r_t) = (1 - \rho) \sum_{j=0}^{\infty} \rho^j (E_t - E_{t-1})(d_{t+1+j}) - \sum_{j=0}^{\infty} \rho^j (E_t - E_{t-1})(r_{t+1+j}). \tag{1.7}$$

The unexpected returns have two components: surprises in future dividends and surprises in future expected returns. Often it is convenient to summarize this expression by the relation

$$r_t - E_{t-1} r_t = \eta_t^d - \eta_t^r. \tag{1.8}$$

These two innovations comprise the news; it is the new information that is used to forecast the discounted future dividends and expected returns.

Each of these innovations is the shock to a weighted average of future dividends or expected returns. Consequently, each is a martingale difference sequence. From (1.7) it is clear that even a small piece of information observed during period t could have a large effect on stock prices if it affects expected dividends for many periods into the future. But it could have a relatively small effect if it only affects dividends for a short period. In the simplest finance world, expected returns are constant, so the second term is zero. However, if there is some predictability in expected returns, either because risk premiums are predictable or because the risk-free rate is changing in a predictable way or because markets are not fully efficient, then the second term may be very important.

The conditional variance of an asset return is simply given from (1.8) as

$$V_{t-1}(r_t) = V_{t-1}(\eta_t^d) + V_{t-1}(\eta_t^r) - 2\operatorname{Cov}(\eta_t^d, \eta_t^r). \tag{1.9}$$

Each term measures the importance of today's news in forecasting the present value of future dividends or expected returns. If d is an infinite-order moving average, possibly with weights that do not converge (like a unit root),

$$d_t = \sum_{i=1}^{\infty} \theta_i \varepsilon_{t-i}^d, \tag{1.10}$$

then

$$\eta_t^d = \varepsilon_t^d (1 - \rho) \sum_{j=0}^{\infty} \rho^j \theta_{j+1} \tag{1.11}$$

and

$$V_{t-1}(\eta_t^d) = V_{t-1}(\varepsilon_t^d) \left[\sum_{j=0}^{\infty} \theta_{j+1} \rho^j (1 - \rho) \right]^2. \tag{1.12}$$

In this model, time variation arises only from substituting volatility in the innovation for dividends. If there is no predictability in expected returns, then this is also the conditional variance of returns. The longer the memory of the dividend process, the more important this effect is and the greater the volatility is. Of course if the dividend process has time variation in the moving-average coefficients, this would be another time-varying component.

The conditional covariance between two asset returns can be expressed in exactly the same terms:

$$\operatorname{Cov}_{t-1}(r_t^1, r_t^2) = \operatorname{Cov}_{t-1}(\eta_t^{d1}, \eta_t^{d2}) + \operatorname{Cov}_{t-1}(\eta_t^{r1}, \eta_t^{r2})$$
$$- \operatorname{Cov}_{t-1}(\eta_t^{d1}, \eta_t^{r2}) - \operatorname{Cov}_{t-1}(\eta_t^{r1}, \eta_t^{d2}). \tag{1.13}$$

In the simple case where expected returns are constant and dividends are fixed-weight moving averages, as in (1.10), and denoting the parameters for each asset as (ρ^1, θ^1) and (ρ^2, θ^2) respectively,

$$\text{Cov}_{t-1}(r_t^1, r_t^2)$$

$$= \text{Cov}_{t-1}(\varepsilon_t^{d1}, \varepsilon_t^{d2})(1 - \rho^1)(1 - \rho^2)\left[\sum_{j=0}^{\infty} \theta_{j+1}^1 (\rho^1)^j\right]\left[\sum_{j=0}^{\infty} \theta_{j+1}^2 (\rho^2)^j\right].$$

$$(1.14)$$

Comparing equations (1.12) and (1.14) makes it clear that the conditional correlation is given simply by

$$\text{corr}_{t-1}(r_t^1, r_t^2) = \text{corr}_{t-1}(\varepsilon_t^{d1}, \varepsilon_t^{d2}). \qquad (1.15)$$

Returns are correlated because the news is correlated. In fact, in this simple case they are equal.

More generally, the relation (1.8) can be used to form a general expression for the covariance matrix of returns. Letting $\boldsymbol{r}$ now represent a vector of asset returns and $\boldsymbol{\eta}$ the vector of innovations due to dividend or expected returns, the equation becomes

$$\boldsymbol{r}_t - E_{t-1}\boldsymbol{r}_t = \boldsymbol{\eta}_t^d - \boldsymbol{\eta}_t^r, \qquad (1.16)$$

where the use of bold emphasizes that these are now vectors. The covariance matrix becomes

$$\text{Cov}_{t-1}(\boldsymbol{r}_t) = \text{Cov}_{t-1}(\boldsymbol{\eta}_t^d) + \text{Cov}_{t-1}(\boldsymbol{\eta}_t^r) - \text{Cov}_{t-1}(\boldsymbol{\eta}_t^d, \boldsymbol{\eta}_t^r) - \text{Cov}_{t-1}(\boldsymbol{\eta}_t^r, \boldsymbol{\eta}_t^d).$$

$$(1.17)$$

Thus correlation will result either from correlation between dividend news events or correlations between risk premiums or expected returns. Most news that is relevant for the future profitability and hence for the dividends of one company will also contain information that is relevant for many other companies. This could be expressed in a factor model, although from the definition of the innovations in (1.8) it is clear that there are many dynamic assumptions implicit in such a representation. Similarly, one might express the covariances of required return in a factor model. Presumably, the factors might include short rates, market risk premiums, credit risk premiums, and possibly other factors. As these are covariances they can easily be more important at some times than at others, so the correlations will sometimes look more like dividend correlations and sometimes more like expected-return correlations. Notice also the cross terms, which could be important but are not well-understood.

Using monthly data Campbell and Ammer (1993) find that the biggest component of the unconditional variance of stock returns is the expected return $V(\eta_t^r)$. Ammer and Mei (1996) find that this is also the biggest component of the correlation between U.K. stocks and U.S. stocks; the covariance between dividend innovations is also significant. The high correlation from news on expected returns makes it critical to understand the source of movements in risk premiums.

1.5 Additional Influences on Correlations

While this analysis has focused on fundamental news as the source of correlations across assets, there are additional considerations that should be mentioned. When returns on two assets are measured over time periods that are not identical, the correlations will be understated. These are called nonsynchronous returns. This affects correlations between assets traded in markets with different trading hours. The correlation between the U.S. market and the Japanese market when measured on a daily closing basis will be much lower than when contemporaneous returns are measured. This is because the closing time in Japan is before the U.S. market opens so some news events will affect the United States one day after they affect Japan. (See Burns et al. (1998), as well as Scholes and Williams (1977) and Lo and MacKinlay (1990a), for a discussion of this and for econometric approaches to the problem.) Burns et al. suggest "synchronizing" the data first. This is applied in Michayluk et al. (2006), where it is demonstrated that synchronized returns on real-estate portfolios are more correlated than they appear using close-to-close returns.

Nonsynchronous returns are at the heart of the late-trading scandal for mutual funds, since late trades allow the investor to observe the news but trade on pre-news prices. To a lesser extent, the same thing happens even with indices that close at last-trade prices for all the components. In this case, some components of the index have stale prices, so the full effect of correlated news will not be seen until the next day. The same effect is present when examining correlations within the day; stale prices will reduce the correlations. Thus a stylized fact is that correlations at high frequencies are often estimated to be smaller than those at low frequencies. This is called the Epps effect after an early paper by Thomas Epps (1979).

Finally, there is much discussion about how correlations between returns can arise through correlated trading or correlated positions. If many portfolios have similar positions, then a news shock to one asset could lead all of the managers to take a similar action with respect to other assets. Such action would lead to correlated order flow and very probably to correlated movements in returns. These correlations might be interpreted as responses to supply and demand effects rather than fundamental news. However, microstructure theory would interpret the order flow as a response to private information that becomes public as trades reveal the information. Thus even correlations that move in response to order flow can be interpreted as being based on news.

In the summer of 1998 following the default of Russian bonds and the decline of the hedge fund Long-Term Capital Management (LTCM), correlations behaved very strangely. It has been argued that many banks and hedge funds held similar positions and as they were unwinding these positions, asset prices of conventionally uncorrelated assets began moving together. Thus trading, and not fundamental news, moved the correlations. Similarly, in August 2007 hedge fund deleveraging is often interpreted as having led to large shifts in correlations. In both cases, however, a more general interpretation is that these trades are informative of the economic conditions in the hedge funds and of the likelihood of future orders. Hence they move prices and correlations.

Internationally, such events are called "contagion." When one emerging market has a financial crisis, often this affects many emerging markets even though they are not economically connected. The link is hypothesized to run through portfolios. It is not clear how important these episodes are to understanding correlations. If there are correlations that are not fundamentally due to news effects, then prices are temporarily or even permanently being driven from their equilibrium value. One could imagine hedge fund strategies to profit from such activities if they were regular and systematic. In fact, hedge funds do play an important role in index reconstitution and other non-information-based trading. Thus contagion effects may well have a basis in news, if only in news about the average investor's tolerance for risk.

Whether trades move markets or whether news moves markets is somewhat of a semantic point. It is generally thought that private information motivates trades and these trades reveal the information. In any case, the concept that news moves asset prices carries with it the idea that the type of news and its intensity will influence correlations. For practical financial decision making, it is necessary to ascertain what types of news are moving the market and forecast how these are likely to evolve in the future.

2
Correlations in Theory

2.1 Conditional Correlations

Correlations measure the linear relationships between two random variables. In this chapter we will discuss a variety of ways to measure correlations in particular settings and will then discuss more general measures of dependence. The standard definition introduced by Pearson emphasizes this linearity. If x and y are random variables, the correlation between them is simply

$$\rho_{x,y} = \frac{E(x - E(x))(y - E(y))}{\sqrt{E(x - E(x))^2 E(y - E(y))^2}}. \tag{2.1}$$

Any such correlation must lie between -1 and 1. This measure of correlation is invariant to univariate linear transformations of the two variables. In particular, the correlation between $x^* = \alpha + \beta x$ and $y^* = \gamma + \delta y$ will be the same as between y and x. The correlation will generally change for nonlinear transformations; consequently, two random variables that are perfectly dependent on each other could still have a correlation less than 1 because they are nonlinearly related. This often happens for financial derivatives such as the return on an option and its underlying asset.

In a time series context, the random variables y and x refer to a specific time period. This could be the return over an interval of time. The expression in this case could depend upon the time at which the random variables are observed. This is called the unconditional correlation and is defined by

$$\rho_{x,y,t} = \frac{E(x_t - E(x_t))(y_t - E(y_t))}{\sqrt{E(x_t - E(x_t))^2 E(y_t - E(y_t))^2}}. \tag{2.2}$$

For this definition to be valid all the moments must exist. In many situations, not only do the means, variances, and covariances all exist, but they are time invariant or covariance stationary. Hence the unconditional correlation is also time invariant. However, the example of asset prices

provides a clear warning. Sometimes investigators examine the correlations between prices rather than those between returns. The unconditional means and variances are generally not defined so it does not make
sense to ask how correlated two asset prices are. It only makes sense to
ask how correlated their returns are. However, even returns may not be
covariance stationary and may have time variation in unconditional correlations. When the euro was launched, the correlations between European equity and bond returns changed permanently. It is probable that
many other movements in correlations as well as volatilities can be
considered to be nonstationary.

A variety of correlations are implied by (2.1) and (2.2) depending
upon the probability density implicit in the expectation. If the density is
the objective density, then these are objective correlations. They could
also be subjective correlations or risk-neutral correlations. In a Bayesian
context they could be prior or posterior correlations.

The density in this expression can also be a conditional density. The
most important conditional density is the time series density conditional
on a previous information set. Thus the conditional correlation of y and
x at time $t + s$ can be expressed as a function of the information available
at time t:

$$\rho_{x,y,t+s/t} = \frac{E_t(x_{t+s} - E_t(x_{t+s}))(y_{t+s} - E_t(y_{t+s}))}{\sqrt{E_t(x_{t+s} - E_t(x_{t+s}))^2 E_t(y_{t+s} - E_t(y_{t+s}))^2}}. \tag{2.3}$$

This is clearly the conditional covariance divided by the product of the
conditional standard deviations. This is a very important conditional correlation for financial applications. Because portfolios must be formed
today based on forecasts of future risks and returns, this correlation is
the one to use. Risks are in the future and hence depend upon future correlations. The number of credit defaults in the future depends upon the
best estimate today. Virtually all financial applications of correlations
involve conditional correlations. When there is no ambiguity about the
conditioning time period the "$/t$" is dropped from the definition. This
will most likely be the case when $s = 1$.

The conditional correlation between asset prices—or, more impor-
tantly, between the log of asset prices—is actually well-defined in many
cases. This is because the conditional mean is simply the lagged asset
price corrected for dividends and expected returns. In most cases, particularly with high-frequency data, the conditional expectation of the log
price is simply the lagged log price, so the terms in the expression for
the conditional correlation are simply returns.

There are, however, other interesting conditioning variables. For some
purposes, it is interesting to ask what the correlation is going to be if

some event happens. What is the correlation between Turkish equities and Greek equities if Turkey joins the European Union? This could only be estimated with a fairly carefully specified model. Another example that is widely used but easily misunderstood is the correlation conditional on an event defined by the two variables. For example, what is the correlation between x and y if both are negative or if both are more negative than a certain threshold? Correlations can be defined for subspaces of the plane but it is not straightforward to interpret the results.

Longin and Solnik (2001) make this point clearly when evaluating the widely accepted hypothesis that correlations are higher when volatilities are higher. For any joint distribution, it is easy to calculate the sample correlation using only data that show large returns. This is then interpreted as a correlation conditional on the event that returns are large. If the original distribution is jointly normal with a positive correlation, then the conditional correlation for large returns will be larger and approaches 1 for the most extreme returns. On the other hand, the correlation conditional on the event that returns are large and positive or large and negative will have correlations that approach 0. These are both properties of the normal distribution rather than evidence of non-normality. Longin and Solnik use extreme value theory to parametrically specify a family of distributions for large returns and conclude that the normal distribution is not a good approximation, at least for the negative tail. They find more dependence in the lower tail than would be expected under normality. However, this result is found under the assumption of time invariance.

Ang and Chen (2002) examine the correlation between stocks and a market index. They find that correlations when the market is declining are much greater than when it is rising. This is especially true for extreme moves. They propose new statistical tests for these differences, which are shown to be significantly different from what would be expected from a multivariate normal distribution.

2.2 Copulas

These correlation measures can be used with any joint density function as long as the relevant moments exist. This is appropriate when the financial planning tool calls for a correlation. It may be that for some questions other measures of dependence would be better, and it may also be the case that for some joint densities more efficient estimators than simply the sample correlation coefficient can be found.

In a multivariate context, the ultimate measure of dependence is the joint density function. The cumulative distribution function (cdf) is defined as

$$F_{1,2,...,n}(y_1,\ldots,y_n) \equiv P(Y_1 < y_1, Y_2 < y_2,\ldots,Y_n < y_n), \qquad (2.4)$$

where the uppercase Ys are the random variables and the lower case ys are a set of real numbers. From this cdf we can derive the univariate cumulative distribution functions as

$$F_i(y_i) = F_{1,2,...,n}(\infty,\infty,\ldots,y_i,\infty,\ldots,\infty) \quad \text{for } i = 1,\ldots,n. \qquad (2.5)$$

Each cdf gives a number between 0 and 1 for any value of y. In particular, we can insert the tth observation on Y_i into its cdf and define the answer to be U_{it}:

$$U_{it} = F_i(Y_{it}), \quad \forall i, t. \qquad (2.6)$$

It was originally shown by Rosenblatt (1952) and more recently used extensively by Diebold et al. (1998, 1999) that these random variables U_{it} are uniformly distributed for each i as long as the original random variables are continuous. That is, the marginal distribution of these Us is uniform. The U_is given by (2.6) will be the same for a random variable as for any linear or nonlinear monotonic transformation of the variable.

Although the Us are marginally uniform, they are not independent. The dependence structure from (2.4) remains. For example, the most extreme occurrences on one variable frequently coincide with the most extreme occurrences of another, suggesting important dependence in the tails. The joint distribution of the Us is called a copula and is given by the expression

$$C(u_1,\ldots,u_n) = F_{1,2,...,n}(F_1^{-1}(u_1),\ldots,F_n^{-1}(u_n)), \qquad (2.7)$$

where the inverses indicate the inverse function. That is,

$$u_1 = F_1(y_1) \quad \Rightarrow \quad y_1 = F_1^{-1}(u_1). \qquad (2.8)$$

If the marginal cdfs are all continuous and strictly monotonic, then the inverses are well-defined and the copula is well-defined for all values on the unit cube. If this is not the case, the copula in (2.7) may not be unique. Similarly, the copula and the marginal distribution functions determine the joint distribution function. This is readily explained by substituting (2.8) into (2.7):

$$F_{1,2,...,n}(y_1,\ldots,y_n) = C(F_1(y_1),\ldots,F_n(y_n)). \qquad (2.9)$$

This result is called Sklar's theorem (see Sklar 1959; McNeil et al. 2005). It gives a general argument for the existence and uniqueness of copulas.

The copula serves a very important role in finance. It summarizes the dependence properties of the data. It gives the probability that multiple assets will all be at their extreme lows or highs at the same time. For risk management purposes, this is a critical issue. Empirically, it certainly appears to be the case that many equity returns are at extreme quantiles at the same time. For example, many equity series had their worst day ever on October 19, 1987. For credit risk problems, a similar situation arises. Firms generally default when their equity value falls to extreme levels. The copula thus indicates the likelihood that multiple firms will all fall to low quantiles at the same time.

The two most widely used copulas are the independent copula and the Gaussian copula. The independent copula is simply given by

$$C_{\text{independent}}(u_1, \ldots, u_n) = \prod_{i=1}^{n} u_i. \tag{2.10}$$

There are no parameters associated with this copula. The Gaussian copula, on the other hand, depends on the correlation matrix. Let Φ^R refer to the multivariate normal cdf with correlation matrix R, where the means are 0 and the standard deviations are 1. Similarly, let Φ be the cdf of a univariate standard normal. The Gaussian copula is then given by

$$C_G^R(u_1, \ldots, u_n) = \Phi^R(\Phi^{-1}(u_1), \ldots, \Phi^{-1}(u_n)). \tag{2.11}$$

Notice that each number between 0 and 1 is translated into a standard normal on the real line. The dependence is then given by the multivariate normal density with covariance matrix R.

If all of these distribution functions are continuously differentiable, then simple expressions for the density functions are available. Using lowercase letters for the joint and univariate density functions,

$$f_{1,2,\ldots,n}(y_1, \ldots, y_n) = \frac{\partial^n F_{1,\ldots,n}(y_1, \ldots, y_n)}{\partial y_1 \partial y_2 \cdots \partial y_n}. \tag{2.12}$$

Applying the same derivative to (2.9) yields

$$f_{1,2,\ldots,n}(y_1, \ldots, y_n) = c(u_1, \ldots, u_n) f_1(y_1) f_2(y_2) \cdots f_n(y_n), \tag{2.13}$$

with the us defined in (2.8). The joint density function is simply the product of all the marginal density functions and the copula density. If these random variables were independent, the joint density would simply be the product of the marginals, so the copula would be the number 1. Notice that this is the result of differentiation of the independent copula in (2.10). Whenever the copula is different from 1, the random variables are no longer independent. The formula is closely related to the familiar moment condition that the covariance between a pair of random variables is simply the product of their two standard deviations and the correlation between them.

Equation (2.13) suggests a mechanism for specifying more general classes of multivariate density functions. One can specify the dependence properties through the copula and the marginal densities through the fs. This is a useful device particularly when the marginals are easy to estimate. When this formulation is used in a conditional setting, both the marginals and the copula can be expressed as conditional densities.

A particularly useful class of joint density functions is based on the Gaussian copula with arbitrary marginal densities. McNeil et al. (2005) call this a meta-Gaussian density. Clearly it becomes a multivariate normal if each of the marginals is also normal, but it can be a fat-tailed or skewed distribution in some or all dimensions and still have a Gaussian copula. Substitution of (2.8) into (2.11) gives the family of meta-Gaussian densities:

$$F(y_1,\ldots,y_n) = \Phi^R(\Phi^{-1}(F_1(y_1)),\ldots,\Phi^{-1}(F_n(y_n))). \qquad (2.14)$$

This family has the property that when each random variable is transformed into a quantile using its marginal distribution and then into a normal using the standard normal inverse cdf, the joint distribution of all of these variables is multivariate normal with a covariance matrix given by the correlation matrix R. It is often convenient to generate such "pseudo-observations," which will be monotonic functions of the original data but which will be multivariate normal with correlation matrix R.

There is now an enormous range of copulas available for applied and theoretical work, although only a few of these copulas are useful in high-dimensional problems. A straightforward generalization of the Gaussian copula is the Student t copula, which has more realistic tail properties. It does not have a closed form but is easily defined from the multivariate Student t distribution.

A class of copulas called the Archimedean class is also useful for expansion to high-dimensional settings. Included in this class are the Gumbel, Clayton, Frank, and generalized Clayton copulas. The class is defined in terms of a copula generator, $\phi(u)$, which is a continuous, convex, and strictly decreasing function defined on the interval $[0,1]$ that ranges from 0 to ∞. The copula is then defined as

$$C(u_1,u_2,\ldots,u_n) = \phi^{-1}(\phi(u_1) + \phi(u_2) + \cdots + \phi(u_n)). \qquad (2.15)$$

For example, the n-dimensional Clayton copula generator is

$$\phi_\theta(u) = \frac{1}{\theta}(u^{-\theta} - 1) \qquad (2.16)$$

and the copula is given by

$$C_\theta^{Cl}(u_1, \ldots, u_n) = (u_1^{-\theta} + \cdots + u_n^{-\theta} - n + 1)^{-1/\theta}, \quad \theta \geq 0. \qquad (2.17)$$

Notice that, unlike the Gaussian and Student t copulas, this n-dimensional joint distribution has only one unknown parameter. It is typically the case that Archimedean copulas have a small number of parameters. The dependence between any pair of random variables is clearly the same. For an extended discussion of these topics and many extensions see McNeil et al. (2005) and the references therein. For a recent survey see Kolev et al. (2006).

Some interesting economic applications develop dynamic or semiparametric copulas. Poon et al. (2004), Jondeau and Rockinger (2006), Patton (2006b), and Bartram et al. (2007) use time-varying copulas to examine nonlinear dependencies. Chen and Fan (2006) and Chen (2007) develop new approaches to estimating dynamic copulas.

2.3 Dependence Measures

Dependence measures for general multivariate densities can now be defined. As pointed out above, correlation is a linear measure of dependence and will not be invariant to nonlinear transformations. It is not difficult to find transformations such that a random variable and its transform are uncorrelated even though they are perfectly dependent. A simple example is the lack of correlation between a symmetric zero-mean random variable and its square. Since the dependence of a set of random variables is measured by the copula, it is natural to have measures that are invariant to the marginal distributions. Any such measures will be invariant to nonlinear monotonic transformations of the data.

The simple Pearson correlation coefficient is clearly not such a measure. It is not determined solely by the copula but also depends upon the marginal densities. However, an invariant measure can be constructed using the meta-Gaussian distribution discussed above. In this case, the *pseudo-correlation* is defined as the Pearson correlation coefficient of the pseudo-observations. Construction of the pseudo-observations requires knowing the marginal distribution and using this to obtain uniform random variables from this distribution and then transforming to normal variables by multiplying by the inverse standard normal distribution. Such pseudo-correlations will be invariant to nonlinear transformations of the original data and will therefore depend solely on the copula. This measure can be expressed as

$$\rho_{ij}^{pseudo} = E(Y_i^* Y_j^*), \quad Y_i^* = \Phi^{-1}(F_i(Y_i)). \qquad (2.18)$$

This measure is called a pseudo-correlation as it is not the correlation of the original data but rather of a transformation of this data. A common assumption on the marginal density is to use the empirical distribution function, which means that the normalized rank is taken as the uniform random variable U and this is then transformed to a set of standard normals.

A closely related measure is the rank correlation or Spearman correlation coefficient. This is defined as the simple correlation coefficient of the quantiles or Us. Since the inputs are invariant to changes in the marginal densities and to monotonic transformations of the data, so is the rank correlation coefficient:

$$\rho_{i,j}^{\text{rank}} = \frac{E(U_i U_j) - 1/4}{1/12}. \tag{2.19}$$

These two measures will not in general be the same but they are usually quite close. If the copula is Gaussian, then they will possibly have the same expectation. If the copula is not Gaussian, then there are settings where the estimates will differ substantially. For example, if the copula has substantial probability near 0 or 1 on both variables, then the pseudo-normals will have observations approaching $\pm\infty$ and will lead to higher pseudo-correlations than rank correlations.

Another widely used measure of dependence is Kendall's tau. To define this measure, think for a moment of two observations on two random variables, (y_1, x_1) and (y_2, x_2). These are called concordant if $y_1 > y_2$ and $x_1 > x_2$ or if $y_1 < y_2$ and $x_1 < x_2$. That is, if they move in the same direction from observation 1 to observation 2. Kendall's tau is the probability that a randomly selected pair of observations will be concordant minus the probability that they will not be:

$$\tau = P(\text{concordant}) - P(\text{not concordant})$$
$$= 2P(\text{concordant}) - 1. \tag{2.20}$$

This measure will clearly be invariant to monotonic transformations of the data and can be computed from the copula directly.

All three of these measures—the pseudo-correlation, the rank correlation coefficient, and Kendall's tau—are nonparametric measures of dependence and consequently they are robust to distributional assumptions and also to outliers. Each of these measures describes the dependence over the full range of outcomes.

Other measures consider the dependence properties for extremes. One measure of extreme dependence is tail dependence. This is a measure of joint extremes. For any particular quantile α you can define the probability that one random variable is beyond this quantile if another is also beyond it. A similar definition can be used for the upper quantiles by reversing the inequalities:

$$\lambda_\alpha = P(Y_1 < F_1^{-1}(\alpha) \mid Y_2 < F_2^{-1}(\alpha))$$
$$= P(U_1 < \alpha \mid U_2 < \alpha)P(U_2 < \alpha \mid U_1 < \alpha). \qquad (2.21)$$

Lower tail dependence is defined as the limit of this probability as the quantile goes to zero:

$$\lambda_{\text{lower}} = \lim_{\alpha \to 0} P(U_1 < \alpha \mid U_2 < \alpha). \qquad (2.22)$$

Upper tail dependence is the corresponding limit as the quantile goes to one. Again, this is a measure that only depends on the quantiles, hence it is a characteristic of a copula.

Lower tail dependence is a measure of particular importance for risk management. Since it is extreme risks that are of most importance, it is necessary to use a copula that appropriately characterizes the probability that multiple assets will have extremely poor returns at the same time. Simple correlations may not accurately reveal the tail dependence. A very important example is the Gaussian copula. For such joint density functions, the tail dependence is always 0 regardless of the correlation structure. Extreme events are always uncorrelated. The multivariate normal thus implies that extreme risks can be diversified away rather easily. It is not a good representation of the tail behavior of asset returns.

Tail dependence is also very important for pricing multiasset credit derivatives and measuring portfolio credit risk. Portfolios with multiple defaultable bonds have increased risk if the probabilities of default are dependent. One default in a diversified portfolio is not serious but many defaults could be disastrous. A company defaults on its debt when the value of outstanding equity becomes very small, as equity holders receive value only after debt holders are paid. Thus the probability that a company will default can be interpreted as the probability that equity values will fall to such a low value. The probability of default can therefore be rephrased as the probability that stock prices fall below a default threshold. For example, if the probability of default within one year is known to be 1%, then the 1% quantile of the end-of-year equity distribution can be interpreted as the default frontier. The default event then becomes the event that equity falls below this quantile. While the probability of this event may be known, the joint distribution of such events is central to models of credit risk. It is straightforward to relate the correlation between default events to the tail dependence of the equity price joint distribution.

Let the default frontier be implicitly defined by the default probability α for two firms. The correlation between the two indicator variables is then the default correlation:

$$
\left.
\begin{aligned}
P(r_{i,t} < d_i) &= \alpha \text{ defines } d, \\
\rho_{i,j}^D &= \mathrm{corr}(I_{\{r_i < d_i\}}, I_{\{r_j < d_j\}}) \\
&= \frac{P(r_{i,t} < d_i \cap r_{j,t} < d_j) - \alpha^2}{\alpha(1 - \alpha)} \\
&= \frac{\lambda_\alpha - \alpha}{1 - \alpha}.
\end{aligned}
\right\}
\tag{2.23}
$$

Thus, for low probability of default where α is small, the default correlation is equal to the lower tail dependence. A central feature of credit risk models is the characterization of these tails. The tail dependence is a function of the copula that is applied.

Recently, an alternative limit approach has been proposed by Ledford and Tawn (1996, 1997, 1998) that parameterizes the rate at which extreme joint probabilities approach 0. The probability that two uniform random variables are below some quantile α can be expressed as

$$
P(U_1 < \alpha, U_2 < \alpha) = \alpha^{2/\phi} L(\alpha),
\tag{2.24}
$$

where $L(\cdot)$ is a slowly varying function with $L(0) = 1$. If the value of ϕ is 1, then the two uniform random variables are asymptotically independent, and if the value of ϕ is 2, then they are perfectly dependent in the limit. Because $P(U < \alpha) = \alpha$, the tail dependence is given by

$$
P(U_1 < \alpha \mid U_2 < \alpha) = L(\alpha)\alpha^{(2-\phi)/\phi}.
\tag{2.25}
$$

This measure of tail dependence will approach 0 for any value of $\phi < 2$, and hence asymptotic independence as measured by tail dependence covers a wide variety of tail probabilities. This approach may prove fruitful in the credit risk problem discussed above.

Resnick (2004) introduces and estimates an extremal dependence measure. This measure is like a correlation weighted to reflect extremes. It is only useful if the tails decline at a particular rate, although other versions could potentially be constructed for other cases.

Various copulas are available for financial applications that have more tail dependence than the Gaussian. The Student t copula, which is defined from the multivariate Student t distribution, has both upper and lower tail dependence. The Clayton and Gumbel copulas can be formulated with either upper or lower tail dependence. Each of these has a single parameter that can be adjusted to match data. For many more copulas and their properties, see Joe (1997) or McNeil et al. (2005).

2.4 On the Value of Accurate Correlations

The risk of holding a portfolio of assets must be evaluated and compared with the expected gains from the position. Without taking risk, there are no excess returns, but not all risks are worth taking. The economic loss associated with an inferior estimate of the covariance matrix of returns results from a faulty evaluation of portfolio risk and consequently a sub-optimal selection of portfolio. This portfolio could have more risk than desired or lower return than desired. In this section, the economic costs of such errors will be calculated to serve as a measure for evaluating estimates of the covariance matrix. This argument is developed in Engle and Colacito (2006). It is useful to restate some familiar results in portfolio theory.

If the portfolio weights are given by the vector w, the vector of N risky returns is r_t, and the risk-free rate is r^f, then the portfolio return is given by

$$\pi_t = \sum_{j=1}^{N} w_j r_{j,t} + \left(1 - \sum_{j=1}^{N} w_j\right) r^f = w'(r_t - r^f) + r^f. \qquad (2.26)$$

Therefore, letting

$$E(r_t - r^f) = \mu, \qquad V(r_t) = \Omega, \qquad (2.27)$$

the portfolio mean and variance are given by

$$E(\pi_t) = w'\mu + r^f, \qquad V(\pi_t) = w'\Omega w. \qquad (2.28)$$

In traditional Markowitz portfolio theory, the risk of a portfolio is the expected variance of the portfolio over a given time period and this is compared with the expected return on the portfolio over the same period. Assuming that the portfolio can be either long or short in all assets, then there are three formulations of the optimal portfolio problem:

$$\min_{w} \quad V(\pi), \qquad (2.29)$$
$$\text{s.t. } E(\pi) \geqslant \mu_0$$

$$\max_{w} \quad E(\pi), \qquad (2.30)$$
$$\text{s.t. } V(\pi) \leqslant k$$

$$\max_{w} E(\pi) - \lambda V(\pi). \qquad (2.31)$$

Either (2.29) or (2.30) can be used to trace out an efficient frontier of portfolios by varying the constraint sets. The optimal choice from this frontier is given directly or indirectly by problem (2.31), which specifies the investor's utility function.

We focus on the first problem, (2.29), which has a solution that depends on the *required excess return*, μ_0:

$$w = \frac{\Omega^{-1}\mu}{\mu\Omega^{-1}\mu}\mu_0. \tag{2.32}$$

The sum of the weights on the risky assets may be less than 1 or greater than 1 depending on whether the investor is long or short in cash. The greater μ_0 is, the less cash is held and the greater the risk that is taken. However, the composition of the portfolio of risky assets is not changed. If assets were independent, then the weights would be greater on assets with higher expected returns and lower variances. With a full covariance matrix, however, this is not precisely true since the correlations matter. When μ_0 is 0, the required return can be achieved with the risk-free rate, so all weights on the risky assets are 0.

If all assets have the same expected return, then the portfolio weights will be selected to minimize variance; optimal portfolios will consist of the minimum-variance portfolio and cash. This is easily seen by substituting $\mu = \iota\mu_1$, where ι is a vector of ones and μ_1 is the excess return expected for all the assets. If $\mu_0 = \mu_1$, then this solution will not involve any cash and will be equivalent to the problem when there is no riskless rate.

The closely related problem of hedging also falls in this framework. If one asset in this universe, say the first, must be held in a portfolio, and the other assets are held simply to reduce risk, then the problem is typically formulated as

$$\min_{\text{s.t. } w_1=1} V(w'r). \tag{2.33}$$

The solution to (2.29) when $\mu = e_1\mu_1$ gives the identical portfolio of risky assets. That is, to find an optimal hedge for asset 1, the investor assumes that only the first asset has a nonzero expected excess return. In general, this optimal hedge will involve some cash position, at least if $\mu_0 < \mu_1$. The wide range of long–short hedges can be analyzed in this way. More sophisticated hedges take account of the expected returns of the assets used in the hedge.

Finally, it is often natural to formulate a tracking error problem by specifying the risk and return relative to a benchmark portfolio. If r^b is the return on the benchmark portfolio, then the problem becomes

$$\max_{w} E(w'(r - r^b)) - \lambda V(w'(r - r^b)). \tag{2.34}$$

Notice that r^b takes exactly the role of the riskless rate. Any investment in the benchmark achieves zero expected excess return and incurs zero risk. Thus the problems formulated above can all be expressed in terms of a benchmark by interpreting μ as the expected return above the benchmark and V as the covariance matrix of returns above the benchmark.

Now consider the portfolio selected by a manager who has incorrect estimates of the covariance of returns and of expected returns. Suppose he believes that the expected return is m and that the covariance matrix of returns is H. The portfolio he will select will be given by (2.32), with H and m replacing their true values. If we call these weights $\hat{w}$, then we obtain

$$\hat{w} = \frac{H^{-1}m}{m'H^{-1}m}\mu_0.$$ (2.35)

This portfolio will not have the mean or variance that he expects. Instead,

$$E(\hat{w}'r) = \frac{\mu'H^{-1}m}{m'H^{-1}m}\mu_0, \qquad V(\hat{w}'r) = \hat{w}'\Omega\hat{w} = \mu_0^2\frac{m'H^{-1}\Omega H^{-1}m}{(m'H^{-1}m)^2}.$$ (2.36)

While the mean and Sharpe ratio could be greater than or less than expected, the variance will always be greater than expected. Engle and Colacito prove that

$$V(\hat{w}'r) - V(w'r) = \mu_0^2\frac{m'H^{-1}\Omega H^{-1}m}{(m'H^{-1}m)^2} - \mu_0^2(m'\Omega^{-1}m)^{-1} \geqslant 0.$$ (2.37)

Thus a measure of the value of good covariance information is the reduction in risk that can be accomplished with a particular vector of expected excess returns, m. A measure of this can also be expressed in terms of the increase in μ_0 that can be achieved with no increase in risk by the manager with accurate covariance information. For different vectors of expected returns, the value of covariance information would in general be different, but it would always be nonnegative. A Bayesian criterion thus averages these losses over the prior distribution of m.

This loss function clearly depends on the true and false covariance matrices in a complex fashion as they appear in inverses and ratios. Engle and Colacito point out that there is a class of errors that has no cost in this context. These are not small errors but they are errors that produce no increase in risk. Analytically, these occur if m is an eigenvector of the matrix ΩH^{-1}. Direct substitution establishes this result. Engle and Colacito have a graphical illustration of this condition showing that it too corresponds to a particular tangency condition.

The result in (2.37) suggests a testing procedure for the accuracy of covariance estimation. Constructing optimal portfolios with a particular vector of excess returns, the portfolio with the smallest variance is the one corresponding to the best covariance estimate. The null hypothesis that these variances are equal can be tested following Diebold and Mariano (2002) for one vector of expected returns. An improved version of the test statistic is given by Engle and Colacito. Similarly, a joint test can be constructed for a finite set of ms.

This approach to distinguishing accurate covariance estimators from inaccurate ones will be used in chapter 9 to compare the different forecasting models introduced in this book. By making simple assumptions on expected returns, minimum-variance portfolios can be constructed that change every time the forecast is updated. The variance of such portfolios is a measure of performance. Two criteria will be used in that section: the first is a minimum-variance portfolio, which is equivalent to assuming that the expected returns are equal; the second is a long–short hedge portfolio, which is equivalent to assuming that one asset has a positive excess expected return while a second asset is merely expected to achieve the riskless rate.

Related measures of performance based on asset allocation or risk management criteria have been used by many authors. Alexander and Barbosa (2008) and Lien et al. (2002) use an out-of-sample hedging criterion to evaluate a covariance forecast. Baillie and Myers (1991) use bivariate models to optimize hedges. Chong (2005) examines forecasting performance of implied volatilities and econometric forecasts. Ferreira and Lopez (2005) and Skintzi and Xanthopoulos-Sisinis (2007) use a Value-at-Risk criterion. Giamouridis and Vrontos (2007) consider optimizing hedge fund performance. Thorp and Milunovich (2007) discuss the benefits of using an asymmetric covariance estimate for asset allocation.

3

Models for Correlation

Since Engle (1982) introduced the idea that volatility could be modeled
and forecast with univariate econometric time series methods, an enormous literature has developed that explores these methods to model
multivariate covariance matrices. Probably the first attempts to estimate
a multivariate GARCH model were by Engle et al. (1984) and by Bollerslev et al. (1988). The literature has been surveyed by Bollerslev et al.
(1994) and by Engle and Mezrich (1996), and more recently by Bauwens
et al. (2006) and by Silvennoinen and Terasvirta (2008). However, even
before these methods were introduced, the practitioner community had
a range of models designed for time-varying correlations based on historical data windows and exponential smoothing. Many of these methods
are widely used today and offer important insights into the key features
needed for a good correlation model.

The papers in this field all seek parameterizations for the covariance
matrix of a set of random variables, conditional on a set of observable
state variables, typically taken to be the past filtration of the dependent
variable. Letting the vector y be a set of random variables such as asset
returns, the objective is to parameterize and estimate

$$H_t = V_{t-1}(y_t). \tag{3.1}$$

From this estimation of the conditional covariances, the conditional correlation and the conditional variances are immediately available. The
standard definition of the conditional correlation between variables i
and j is

$$\rho_{i,j,t} = \frac{E_{t-1}((y_{i,t} - E_{t-1}(y_{i,t}))(y_{j,t} - E_{t-1}(y_{j,t})))}{\sqrt{V_{t-1}(y_{i,t})V_{t-1}(y_{j,t})}} = \frac{H_{i,j,t}}{\sqrt{H_{i,i,t}H_{j,j,t}}}. \tag{3.2}$$

In matrix notation, the conditional correlation matrix and the variance
matrix are given by

$$R_t = D_t^{-1}H_tD_t^{-1} \quad \text{and} \quad D_t^2 = \text{diag}[H_t], \tag{3.3}$$

where the notation diag[A] refers to a matrix that has the same diagonal elements as A but has all other elements 0. These expressions imply the familiar representation for a covariance matrix:

$$H_t = D_t R_t D_t. \tag{3.4}$$

Both H and R will clearly be positive definite if there are no linear dependencies in the random variables. The parameterization for this dynamic problem should insure that all covariance and correlation matrices are positive definite and hence that all volatilities are positive. Since these matrices are actually stochastic processes, they are required only to be positive definite with probability 1. That is, for all past histories with positive probability, the covariance matrix should be positive definite. If this is not the case, then there will be linear combinations of y that apparently have zero or negative variances. The conditional variance of a portfolio with weights w is

$$V_{t-1}(w'y_t) = w'H_t w, \tag{3.5}$$

which will be positive if H is positive definite. If H is only positive semidefinite, then there will be zero-variance portfolios; and if it is indefinite, there will be negative variance portfolios as well. In asset allocation problems or risk management problems, portfolios are optimized to reduce risk. Clever algorithms will find these apparently riskless or negative-risk portfolios and put capital into them; in order for these applications to be successful it is essential to eliminate this possibility. Generally, the existence of negative- or zero-variance portfolios must be considered to be a misspecification of the covariance matrix.

3.1   The Moving Average and the Exponential Smoother

The most widely used covariance matrix estimators are the simplest. These treat each element of the covariance matrix the same and often assume that the mean is 0. The models are the moving-average volatilities and correlations, often called *historical volatilities and correlations*:

$$H_{i,j,t}^{hist} = \frac{1}{m} \sum_{s=t-m}^{t-1} y_{i,s} y_{j,s} \quad \text{for all } i, j. \tag{3.6}$$

The *exponential smoother* has been used by RiskMetrics for its risk calculations. It is defined by

$$H_{i,j,t}^{ex} = \lambda y_{i,t-1} y_{j,t-1} + (1-\lambda) H_{i,j,t-1} \quad \text{for all } i, j$$

$$= \lambda \sum_{s=1}^{\infty} (1-\lambda)^s y_{i,t-s-1} y_{j,t-s-1}. \tag{3.7}$$

In both of these models the covariance matrix for the observation at time t is based on information through time $t - 1$. In each case there is a single parameter that governs the estimation of the entire covariance matrix: m in the moving-average model and λ for the exponential smoother.

These covariance estimators will be positive definite under weak assumptions. The conditions are easier to see in matrix representation. Let y_t be the $n \times 1$ vector of asset returns, then the estimators can be written

$$H_t^{\text{hist}} = \frac{1}{m} \sum_{s=t-m}^{t-1} y_s y_s', \qquad H_t^{\text{ex}} = \lambda y_{t-1} y_{t-1}' + (1 - \lambda) H_{t-1}^{\text{ex}}. \qquad (3.8)$$

Each estimator is the average or the weighted average of rank 1 positive semidefinite matrices and consequently is at least positive semidefinite. The exponential smoother will be positive definite if the initial H_1 is positive definite, but in general the positive definiteness requires very weak assumptions on the ys, such as that they have a nonsingular covariance matrix and are only weakly dependent. Of course, the historical covariance matrix cannot be positive definite unless $m > n$.

Each of these estimators has a "start-up" problem that requires assumptions on distribution of the ys before the beginning of the sample. The simplest solution and one that is widely used in theoretical studies is that the pre-sample values of y are all 0. In this case, the moving-average model will ramp up for the first m days and will be of reduced rank for this period. For the exponential smoother, this will have a gradually decreasing effect throughout the data set. Empirically, this is not very attractive. A second assumption is that the first $m > n$ days are used to estimate the starting covariance matrix. The first estimate is not until date m. This is the most popular solution for the moving average but is rarely used for the exponential smoother. A third solution is to assume that the covariance matrix of the pre-sample values is the same as the unconditional covariance matrix of the full data. In this case the sample covariance matrix of the full data set is used to initialize the model. This is commonly used for multivariate GARCH models and for exponential smoothers. Even this assumption may be unattractive in some cases, however. Suppose there are trends in the data, then the sample average will be different from the pre-sample average and there will be an initial adjustment. It is possible to forecast backward, or "backcast," these values. This is not used in any models I know of, but it is implemented in some univariate volatility models.

The unknown parameters in these two models are often not estimated but are simply assumed based on the experience of the investigator. For example, RiskMetrics chooses $\lambda = 0.06$ for all assets for daily data, and on Wall Street many covariances are called "historical" and are based on twenty-day or hundred-day moving averages. These models are very simple and have some serious shortcomings, but for some tasks they perform quite well.

3.2 Vector GARCH

A wide range of multivariate GARCH models have been proposed and used. These have been surveyed by several authors—including Bollerslev et al. (1994) and, more recently, Bauwens et al. (2006) and Silvennoinen and Terasvirta (2008)—and will not be discussed in great detail here. It is sufficient here to outline how they are typically formulated. The models can be described by an equation for each element of the covariance matrix. The most popular class, called *diagonal multivariate GARCH* or *diagonal vector GARCH*, formulates the i, j element of the covariance matrix in terms of only the product of the prior i and j returns. For example, a first-order version of the diagonal vector GARCH is

$$h_{i,j,t} = \omega_{i,j} + \alpha_{i,j} y_{i,t-1} y_{j,t-1} + \beta_{i,j} h_{i,j,t-1}. \tag{3.9}$$

Each element of the covariance matrix is a function only of its own past information set. This model obviously has a great many parameters and will not in general produce positive-definite covariance matrices. However, with restrictions on the parameters, positive-definite symmetric covariance matrices can be guaranteed. A popular set of restrictions is called the *diagonal BEKK* and is defined by

$$\alpha_{i,j} = \alpha_i \alpha_j, \qquad \beta_{i,j} = \beta_i \beta_j. \tag{3.10}$$

A simple restriction forces all the αs to be the same and all the βs to be the same and then requires the ω matrix to be positive definite. This is often called the first-order *scalar diagonal multivariate GARCH* or *scalar MGARCH*:

$$h_{i,j,t} = \omega_{i,j} + \alpha y_{i,t-1} y_{j,t-1} + \beta h_{i,j,t-1}. \tag{3.11}$$

The model has $\frac{1}{2} n(n-1) + 2$ unknown parameters, all but two of which are in the intercept.

In the opposite direction, much more complex models can be described. These make each element of the covariance matrix a linear function of the squares and cross products of the data. This is the *vec* model, which in first order has typical element

$$h_{i,j,t} = \omega_{i,j} + \sum_{k,l=1}^{n} (\alpha_{i,j,k,l} y_{k,t-1} y_{l,t-1} + \beta_{i,j,k,l} h_{k,l,t-1}). \tag{3.12}$$

Since it includes squares and cross products of data and past conditional variances other than the i, j elements, it has vastly more parameters and without a great many restrictions, it will not necessarily generate positive definitene covariance matrices. This model is too general to be useful but many special cases are used.

Although the vec model has very generous parameterization, it is still linear in the squares and cross products of the data, which is of course a severe restriction. This is convenient from a theoretical perspective as it supports multistep analytical forecasting, but it is not necessarily consistent with the evolution of covariance matrices. This linearity is abandoned in the models discussed in sections 3.4 and 3.6.

3.3 Matrix Formulations and Results for Vector GARCH

This section is more mathematically sophisticated than the previous one and can be skipped without loss of continuity. Here the interested reader can find a more detailed analysis of the vec representation and its special cases. This follows much of Engle and Kroner (1995). A simple description of this class of models is that all the variances and covariances depend upon all the squares and cross products of the data. This is summarized using the vec notation. The vec operation converts a matrix to a vector by stacking all the columns of the matrix into one very long vector. If A is an $n \times n$ matrix, then $\text{vec}(A)$ is an $n^2 \times 1$ vector. If A is symmetric, then there will be a lot of duplication in $\text{vec}(A)$ as only about half the elements are unique. The vec model can be written as

$$\text{vec}(H_t) = \text{vec}(\Omega) + \sum_{s=1}^{p} A_s^* \, \text{vec}(y_{t-s} y_{t-s}') + \sum_{s=1}^{q} B_s^* \, \text{vec}(H_{t-s}). \quad (3.13)$$

This model describes the dependence between each element of H and squares and cross products of past returns and lagged covariances.

Because of the linearity of this system, it is quite easy to forecast and check stationarity. Calculating a k-steps-ahead forecast of the covariance of the ys simply requires recursive evaluation. If $k > p, q$, then

$$E_t(\text{vec}(y_{t+k} y_{t+k}')) = E_t(\text{vec}(H_{t+k}))$$

$$= \text{vec}(\Omega) + \sum_{s=1}^{p} (A_s^* + B_s^*) E_t(\text{vec}(H_{t+k-s})). \quad (3.14)$$

Both equalities apply iterative expectations, recognizing that $E_t(\cdot) = E_t E_{t+k-s}(\cdot)$.

As the forecast horizon goes to infinity, the forecasts will mean revert to a fixed-variance covariance matrix if the process is covariance stationary. This condition depends upon the eigenvalues of the difference equation (3.14). Engle and Kroner (1995) prove the following theorem.

Theorem 3.1. *If (3.1) and (3.13) define the stochastic process of a multivariate conditional covariance matrix with parameters that guarantee positive-definite covariances and if*

$$\left| I - \sum_{s=1}^{p} (A_s^* + B_s^*) z^s \right| = 0 \tag{3.15}$$

has all of its solutions outside the unit circle, then $\{y_t\}$ *is a covariance-stationary process with unconditional covariance matrix*

$$E(\text{vec}(y_t y_t')) = \left[I - \sum_{s=1}^{p} (A_s^* + B_s^*) \right]^{-1} \text{vec}(\Omega). \tag{3.16}$$

Proof. See Engle and Kroner (1995, p. 133).   □

The vec model potentially has an enormous number of free parameters. The intercept has n^2 parameters, of which about half are free since it is a symmetric matrix. The coefficient matrices have n^4 parameters; through symmetry only about a quarter of these are free. Thus there are approximately $\frac{1}{4} n^4 (p + q + 2)$ parameters. For a $(1, 1)$ model with five assets there would be approximately 625 parameters, and for a ten-asset model, approximately 10,000 parameters. This is clearly unmanageable. Furthermore, this model is not guaranteed to be positive definite. Thus it is natural to attempt to tighten this parameterization and from this comes the biggest class of multivariate GARCH models.

The moving average and the exponential smoother are extremes in the restricted parameterization of this model. If $p = q = 1$, if A and B are scalars that add up to 1, and if the intercept is 0, then this is the exponential smoother. If $p = m$, $q = 0$, and all the A matrices are simply the scalar $1/m$ and again the intercept is 0, then we get the moving-average model. Testing these restrictions against more general representations can therefore assess the effectiveness of these simple models.

A less restrictive class is the diagonal vec models. In this model the parameter matrices A and B are assumed to be diagonal. This is a natural restriction. It means that each covariance depends only on past values of the cross products of its own returns and on its past conditional covariance. Thus the relevant covariance element is some weighted average of past cross products, where the weighting reflects the speed of information decay.

The model can then be described by

$$H_{i,j,t} = \omega_{i,j} + \sum_{s=1}^{p} a_{i,j,s} y_{i,t-s} y_{j,t-s} + \sum_{s=1}^{q} b_{i,j,s} H_{i,j,t-s}. \tag{3.17}$$

This model has approximately $\frac{1}{2}(p+q+1)n^2$ parameters. It does not have any guarantee of positive definiteness however. To establish conditions for the diagonal vec multivariate GARCH model to be positive definite, we turn to a theorem used by Ding and Engle (2001).

Multiplication of two matrices of identical dimension can be defined as an element-by-element product, or Hadamard product, using the symbol "$\odot$":

$$[A \odot B]_{i,j} = A_{i,j}^* B_{i,j}. \tag{3.18}$$

The Hadamard product has several useful properties and these are incorporated in the following two lemmas from Ding and Engle (2001).

Lemma 3.2. *If A is an $n \times n$ symmetric positive-definite matrix and b is an $n \times 1$ nonzero vector, then $C = A \odot bb'$ is positive definite, and if A is positive semidefinite, then C is positive semidefinite.*

The proof simply rewrites this product as $C = \text{diag}(b) A \text{diag}(b)$, where $\text{diag}(b)$ is a matrix with b on the diagonal. The conclusion follows immediately.

Lemma 3.3. *If A and B are positive semidefinite symmetric matrices, then $C = A \odot B$ is also positive semidefinite.*

The proof rewrites B in its spectral representation and recognizes that all eigenvalues must be nonnegative and real. Thus,

$$C = A \odot \sum_{i=1}^{n} \lambda_i b_i b_i' = \sum_{i=1}^{n} \lambda_i (A \odot b_i b_i'), \tag{3.19}$$

which is a weighted sum of positive semidefinite matrices. Styan (1973) gives more details on Hadamard products.

With this tool, the diagonal vec model in (3.17) can be rewritten as

$$H_t = \Omega + \sum_{s=1}^{p} A_s \odot (y_{t-s} y_{t-s}') + \sum_{s=1}^{q} B_s \odot H_{t-s}, \tag{3.20}$$

where the A and B matrices are the collected values of the as and bs in (3.17), which were themselves all diagonal elements of A^* and B^* in (3.13). The two lemmas can now be used to establish parameter restrictions for positive definiteness.

Lemma 3.4. *If all the A and B matrices are positive semidefinite and if Ω is positive definite, then H will be positive definite for any t.*

The proof again follows from applying the two lemmas to the sums of matrices that make up H_t.

Ding and Engle then describe several parameterizations that will ensure that positive definiteness is satisfied. In each case the intercept matrix must be positive definite. The simplest is the *scalar-diagonal* model, in which each matrix is simply one parameter times a matrix of ones.

Scalar-diagonal: $A_s = \alpha_s \iota \iota'$, $B_s = \beta_s \iota \iota'$, where α and β are scalars and ι is a vector of ones.

The second is a vector outer product.

Vector-diagonal: $A_s = a_s a'_s$, $B_s = b_s b'_s$, where a and b are $n \times 1$ vectors.

And, finally, the matrix outer product is given as follows.

Matrix-diagonal: $A_s = \Psi_s \Psi'_s$, $B_s = Y_s Y'_s$, where Ψ and Y are matrices, possibly with reduced rank.

Clearly, these models all satisfy the assumptions of lemma 3.4. Furthermore, combinations of these models will also satisfy lemma 3.4. Thus not all lags need to have the same form and the ARCH and GARCH terms can be of different types.

Another general class of the vec models that guarantees positive definiteness can be established. These were proposed by Engle and Kroner (1995) and are named for a series of graduate students who worked on these models: Yoshi Baba, and Dennis Kraft, along with Ken Kroner and myself. The BEKK representation is given by

$$H_t = \Omega + \sum_{s=1}^{p} \sum_{k=1}^{K} \Phi'_{k,s} y_{t-s} y'_{t-s} \Phi_{k,s} + \sum_{s=1}^{q} \sum_{k=1}^{K} \Theta'_{k,s} H_{t-s} \Theta_{k,s}, \qquad (3.21)$$

which is clearly positive definite if Ω is positive definite. The matrices of coefficients here can be full, symmetric, reduced rank, diagonal, or can even be scalars times the identity. None of this will affect the positive definiteness or symmetry of the matrix H. To see the relation between this model and the others proposed above, we need lemma 3.5.

Lemma 3.5. *For any three conformable matrices A, B and C, $\text{vec}(ABC) = A \otimes C' \text{vec}(B)$, where the symbol "$\otimes$" refers to a tensor product.*

Taking the vec of both sides of (3.21) gives

$$\text{vec}(H_t) = \text{vec}(\Omega) + \sum_{s=1}^{p} (\Phi_s \otimes \Phi_s) \text{vec}(y_{t-s} y'_{t-s}) + \sum_{s=1}^{q} (\Theta_s \otimes \Theta_s) \text{vec}(H_{t-s}),$$

$$(3.22)$$

which is clearly a vec model as in (3.13). The big coefficient matrices in the vec model have n^2 coefficients rather than roughly $\frac{1}{4}n^4$. These models correspond to nondiagonal versions of the vec model that are positive definite. The nondiagonal structure allows squares and cross products of one asset to help predict variances and covariances of other assets. It seems to be important to allow for this possibility. In fact, however, there are few striking examples of this in the literature.

3.4 Constant Conditional Correlation

Another class of multivariate GARCH models was introduced by Bollerslev (1990) and is called constant conditional correlation, or CCC. In this model the conditional correlations between each pair of assets are restricted to be time invariant. Thus the covariance matrix is defined by

$$H_{i,j,t} = \rho_{i,j}\sqrt{H_{i,i,t}H_{j,j,t}}, \tag{3.23}$$

or, in matrix notation,

$$H_t = D_t R D_t, \qquad D_t = \mathrm{diag}(H_t)^{1/2}. \tag{3.24}$$

The variances of the ys can follow any type of process, but the conditional covariances must move just enough to keep the correlations constant.

This model is not linear in squares and cross products of the data. Multistep forecasts of the covariance matrix cannot be made analytically, although approximations are available. Similarly, the unconditional covariance matrix cannot be calculated exactly. This model is the most tractable model to use for large systems. It has a well-defined likelihood function and is easily estimated with a two-step approach. The first step involves estimating univariate models and the second step simply computes the sample correlation between the standardized residuals.

The computational appeal of this estimator is offset by the obvious drawback that we cannot learn about changes in correlations while assuming they are constant. It may be that in some periods the conditional correlations between some variables are constant. However, this can only be a consequence of extensive testing, not an a priori assumption. More flexible models are surely needed.

3.5 Orthogonal GARCH

One generalization of this model assumes that a nonsingular linear combination of the variables has a CCC structure. Most often, the assumption is also made that this correlation matrix is the identity, $R = I$. Suppose there is a matrix P with the property that

$$V_{t-1}(Py_t) = D_t R D_t, \qquad V_{t-1}(y_t) = P^{-1} D_t R D_t P'^{-1}. \qquad (3.25)$$

It is clear that in this model, y is not a CCC model, even though Py is. The conditional correlation matrix is given by

$$\text{corr}_{t-1}(y_t) = \text{diag}(P^{-1} D_t R D_t P'^{-1})^{-1/2} P^{-1} D_t R D_t P'^{-1}$$
$$\times \text{diag}(P^{-1} D_t R D_t P'^{-1})^{-1/2}, \qquad (3.26)$$

which is time varying. If the matrices P and D commute, then this matrix will no longer depend on D_t so it will be time invariant. In general, P and D will not commute unless P is itself diagonal. Hence (3.25) is indeed a generalization with the parameters of P potentially estimable.

A natural version of this estimator is to specify that P is triangular. Without loss of generality, the elements of the P matrix can be chosen to make the unconditional covariance matrix of the linear combinations diagonal. Consequently, the unconditional correlation matrix is the identity matrix. This is the Cholesky factorization of the covariance matrix. It is most easily computed by least-squares regressions. That is, regress y_2 on y_1, and y_3 on y_1 and y_2, and so forth. The residuals will be uncorrelated across equations by construction. This model is now expressed more concisely as

$$V(Py_t) = \Lambda, \qquad \Lambda \sim \text{diagonal}. \qquad (3.27)$$

The assumption that is then made is that the conditional covariance matrix is diagonal with univariate GARCH for each series. This assumption is the heart of the method. It is equivalent to assuming that this linear combination of the variables is a CCC model as in (3.25). Because the unconditional covariance matrix is diagonal, it follows that $R = I$. Each of the residual series computed in (3.27) is then taken to be a GARCH process and its conditional variance is estimated. Mathematically this can be written as

$$\left. \begin{array}{l} V_{t-1}(Py_t) = G_t^2, \quad G_t = \text{diag}(\sqrt{h_{1,t}}, \sqrt{h_{2,t}}, \dots, \sqrt{h_{n,t}}), \\ h_{i,t} \sim \text{GARCH}, \quad i = 1, \dots, n. \end{array} \right\} \qquad (3.28)$$

The final covariance matrix is then reconstructed as

$$V_{t-1}(y_t) = P^{-1} G_t^2 P'^{-1}. \qquad (3.29)$$

The version of this model that is widely used and that has been popularized by Alexander (2002) and Alexander and Barbosa (2008) is called OGARCH, for orthogonal GARCH. As before, it is assumed that each of the diagonal conditional variances is a univariate GARCH model. This method is formally the same as (3.28) and (3.29) but corresponds to a different choice of P. In this case P^{-1} is the matrix of eigenvectors of the unconditional covariance matrix. The random variables Py are then called the *principal components* of y:

$$V(y_t) = \Sigma, \quad \Sigma = P^{-1}\Lambda P'^{-1}, \quad \Lambda \sim \text{diagonal}. \tag{3.30}$$

Assuming that the conditional covariance matrix is simply given by

$$V_{t-1}(Py_t) = G_t^2, \quad V_{t-1}(y_t) = P^{-1}G_t^2 P'^{-1}, \tag{3.31}$$

with each component following a univariate GARCH. The unconditional covariance matrix is

$$V(y_t) = P^{-1}E(G_t^2)P'^{-1} = P^{-1}\Lambda P'^{-1} = \Sigma. \tag{3.32}$$

A closely related choice, favored by Alexander, first transforms y to have unit variance so that the eigenvectors and principal components are now computed from the unconditional correlation matrix. Without loss of generality, the eigenvectors of the correlation matrix can be found and expressed as $\tilde{P}^{-1}$. That is,

$$\tilde{D} \equiv \text{diag}(\Sigma)^{1/2}, \quad V(\tilde{D}^{-1}y_t) \equiv \tilde{R} = \tilde{P}^{-1}\tilde{\Lambda}\tilde{P}^{-1}. \tag{3.33}$$

The OGARCH model then assumes that

$$V_{t-1}(\tilde{P}\tilde{D}^{-1}y_t) = \tilde{G}_t^2, \quad \tilde{G}_t \text{ is diagonal}, \tag{3.34}$$

with components that are univariate GARCH models. Hence the conditional covariance matrix is given by

$$V_{t-1}(y_t) = \tilde{D}\tilde{P}^{-1}\tilde{G}_t^2\tilde{P}'^{-1}\tilde{D}. \tag{3.35}$$

Once P or $\tilde{P}$ and $\tilde{D}$ are computed from the data, the variances of each of the principal components can be estimated according to (3.31) or (3.34). This is a two-step estimator: first extract the principal components from S and then estimate univariate models for each of these. The econometric analysis of this two-step process has not yet been examined.

These choices of P lead to different models. In fact there are many choices. Van der Weide (2002) has recently recognized this feature and introduced the class of generalized orthogonal GARCH, or GO-GARCH.

3.6 Dynamic Conditional Correlation

A natural generalization of the CCC model is the dynamic conditional correlation (DCC) model of Engle (2002a). It follows the same structure as the CCC model but allows the correlations to vary rather than requiring them to be constant. This model formulates the volatilities of a set of random variables in one set of equations and the correlations between them in another set. Correlations can be interpreted as a stochastic process that can be specified without concern for the volatility process. The reason this is possible can be easily seen from the definition of a conditional correlation and the algebra in (3.4).

If $\{y_1, y_2\}$ are two zero-mean random variables, then the conditional correlation between them is defined in (3.2). Recognizing that these random variables can be written as the product of their conditional standard deviation and an innovation,

$$y_{1,t} = \sqrt{E_{t-1}(y_{1,t}^2)}\,\varepsilon_{1,t}, \qquad y_{2,t} = \sqrt{E_{t-1}(y_{2,t}^2)}\,\varepsilon_{2,t}, \qquad (3.36)$$

the conditional correlation is immediately seen to be equal to

$$\rho_{1,2,t} = \frac{E_{t-1}(\varepsilon_{1,t}\varepsilon_{2,t})}{\sqrt{E_{t-1}(\varepsilon_{1,t})E_{t-1}(\varepsilon_{2,t})}} = E_{t-1}(\varepsilon_{1,t}\varepsilon_{2,t}). \qquad (3.37)$$

Hence these shocks play an important role in correlation estimation as they contain all the relevant information for estimating correlations. These are called standardized residuals, or volatility-adjusted returns. They must have a conditional variance of 1 as well as an unconditional variance of 1. They have conditional and unconditional means of 0. In each case the information set for these conditional moments is the joint past. In some cases the univariate variance models must be formulated with multivariate information sets. The cross products of these standardized residuals may have nonzero means that can be predicted by other functions of the data. It is this predictability that can be modeled to give time-varying conditional correlations. In the CCC model there is no predictability and consequently the conditional correlations are constant.

Rather than estimating the covariance matrix and then calculating the conditional correlations from it, the DCC model uses the standardized residuals and estimates the correlation matrix directly. This gives enormous flexibility in the model. Many different parameterizations have been proposed but they all have a similar objective. In each period the new information on the volatility-adjusted returns is used to update the correlation estimates. When returns are moving in the same direction the correlations should be increased, and when they are moving in the opposite direction correlations should be decreased. This process may be mean reverting to typical values of the correlations, in the sense that high correlations will tend to fall and low correlations will rise. It may be augmented by some economic events, such as down markets or macroeconomic variables.

In the rest of the book this model will be examined for its econometric implementation and its economic implications. There are many extensions that are usefully employed in certain circumstances. The applicability of this model to large-dimension problems is of great interest and its ability to anticipate correlations will be examined.

3.7 Alternative Approaches and Expanded Data Sets

The rest of this book will focus on the DCC model and some of its most important extensions. This survey of methods is, however, still far from complete. There are many interesting new approaches being developed to model correlations. Some use new flexible functional forms. Others are based on new data sets, particularly those using intradaily data. These include models based on daily range statistics and realized volatilities and correlations.

Niguez and Rubia (2006) show how long-memory volatility processes can be generalized to the multivariate context. Asai et al. (2006) survey the applications of multivariate stochastic volatility models. Chib et al. (2006) and Yu and Meyer (2006) demonstrate feasibility for large systems. Andersen et al. (2005) discuss methods using realized volatilities and correlations. Hafner et al. (2006) introduce a semiparametric specification. He and Terasvirta (2004) consider a class of multivariate GARCH models that generalizes the constant conditional correlation model. Audrino and Barone-Adesi (2006), Ledoit et al. (2003), Ledoit and Wolf (2003), and Dellaportas and Vrontos (2007) introduce models that combine static and dynamic models to give, essentially, a shrinkage style estimator. Bekaert and Harvey (1995) and Gallo and Otranto (2007) examine regime-switching models of correlation. Braun et al. (1995) build multivariate models based on the exponential GARCH (EGARCH) specification and a companion model for conditional betas. Storti (2008) introduces a multivariate version of their bilinear GARCH. Kawakatsu (2006) and Caporin (2007) introduce an exponential model in which variances have predictable effects on correlations. Christodoulakis (2007) parameterizes the Cholesky factorization of the covariance matrix. Palandri (2005) decomposes a correlation matrix into a collection of partial correlations which can be parameterized. McAleer and da Veiga (2008) model covariances with what they call a "spillover" effect. Rosenow (2008) shows how random matrix theory can be used to understand the sampling distribution of correlation matrices and estimate the optimal number of principal components and new classes of multivariate GARCH models. Harris et al. (2007) build simple univariate models of portfolios in order to infer the full covariance matrix.

4

Dynamic Conditional Correlation

There are three general steps in the specification and estimation of a DCC model. First, the volatilities must be estimated to construct standardized residuals or volatility-adjusted returns. This is often called "DE-GARCHING" the data. Then the quasi-correlations must be estimated in a dynamic fashion based on these standardized residuals. Finally, the estimated quasi-correlation matrix must be rescaled to ensure that it is a proper correlation matrix. These three steps will be considered in turn in this chapter. The estimation strategy will then be considered as a whole.

4.1 DE-GARCHING

The DCC model introduced in the previous chapter formulated the covariance matrix of a vector of returns in the familiar form of a standard deviation matrix and a correlation matrix:

$$E_{t-1}(y_t y_t') = H_t = D_t R_t D_t, \qquad D_t^2 = \text{diag}\{H_t\}. \qquad (4.1)$$

In matrix notation, it follows that the conditional correlation matrix is simply the covariance matrix of the standardized residuals:

$$R_t = V_{t-1}(D_t^{-1} y_t) = V_{t-1}(\varepsilon_t). \qquad (4.2)$$

The information that is needed to estimate the conditional correlations is summarized in these volatility-adjusted returns. The problem of estimating the elements in H can be separated into the problem of estimating the diagonal elements and then using these to estimate the off-diagonal elements.

This step should be considered more carefully as it is central to this whole class of estimators. The diagonal elements of D_t are the square root of the expected variance of each asset based on the past information set. That is,

$$H_{i,i,t} = E_{t-1}(y_{i,t}^2). \qquad (4.3)$$

This is a problem that has received a great deal of attention. What is the variance of a random variable given past information? A simple answer is presented by ARCH/GARCH models.

For example, the GARCH$(1, 1)$ model defines this as

$$H_{i,i,t} = \omega_i + \alpha_i y_{t-1}^2 + \beta_i H_{i,i,t-1}. \tag{4.4}$$

This model can be estimated for each asset separately to get its conditional variance, and then the standardized residuals are defined by

$$\varepsilon_{i,t} = y_{i,t} / \sqrt{H_{i,i,t}}. \tag{4.5}$$

This operation can be done using conventional software quite quickly and converts data with time-varying volatility to data with unit volatility. When plotted, it should have constant amplitude and it should not have significant autocorrelation in the squared returns. This process is sometime called "DE-GARCHING" a series.

The basic model that gives its name to the procedure is the GARCH model of Bollerslev (1986). In most of the analyses in this book, an asymmetric version of the GARCH model is used. This idea was introduced by Nelson (1991), who formulated the exponential GARCH, or EGARCH, model. We will use the threshold ARCH, or TARCH, model introduced by Glosten et al. (1993) and by Rabemananjara and Zakoian (1993). An economic model that explains why asymmetric volatility is so commonly observed in equity data was introduced and then carefully modeled by French et al. (1987), Campbell and Hentschel (1992), and Bekaert and Wu (2000), and it was then applied to bond returns by de Goeij and Marquering (2006). The volatility specification can be expressed as

$$H_{i,i,t} = \omega_i + \alpha_i y_{t-1}^2 + \gamma_i y_{t-1}^2 I_{y_{t-1}<0} + \beta_i H_{i,i,t-1}. \tag{4.6}$$

In this model, the impact of negative returns can be different from the impact of positive returns. In many equity series, and in particular for market indices, negative returns are much more influential than positive returns.

The flexibility of this approach to DE-GARCHING is worth mentioning. The parameters for different assets can be different. A wide range of GARCH-style models can be used and there is no need to use the same model for each series. For example, asymmetric volatility models such as TARCH, EGARCH, and PARCH can be used when the data warrant. Nonstationary models such as the Spline GARCH of Engle and Rangel (2008) or the filtering methods of Nelson and Foster (1994) can be used. The model can also include predetermined regressors in either the mean or the variance. Some interesting examples are given in the papers by

Chou (2005), Engle (2002b), and Fernandes et al. (2005), who use the daily range to estimate the volatilities in this framework. Furthermore, the model could in principle be estimated using a stochastic volatility model following any of the many different approaches that have been tried. It can even be estimated using measures such as implied volatilities or variance swap rates if these are available for some of the series. There are many sources for reading about GARCH and related models. Many of them are listed in Engle (1995), Bollerslev et al. (1992, 1994), Engle (2001), and Engle and Patton (2001) and are classified in a glossary by Bollerslev (forthcoming).

It is also possible to DE-GARCH many series jointly. This may be sensible since volatilities are observed to move together. Anderson and Vahid (2007), Bauwens and Rombouts (2007), and Engle and Marcucci (2006) investigate multiasset volatility models, which can benefit from common elements in their stochastic processes. Even volatility spillovers like those in Engle et al. (1990a) and Milunovich and Thorp (2006) can be usefully introduced here.

4.2 Estimating the Quasi-Correlations

The three most commonly used specifications for the quasi-correlation matrix of the standardized returns are the integrated, mean-reverting, and asymmetric models. In all cases, a stochastic process is proposed for a matrix Q that is an approximation to the correlation matrix. We will call it the quasi-correlation matrix.

In the *integrated* model, this stochastic process Q is calculated between any two variables as

$$Q_{1,2,t} = \lambda \varepsilon_{1,t-1} \varepsilon_{2,t-1} + (1 - \lambda)Q_{1,2,t-1}. \tag{4.7}$$

This is a direct analogue of exponential smoothing except that the data are volatility adjusted. There is only one parameter, λ, and this is used in each equation of the system. This model is called the *integrated* DCC model because the process for Q has a unit root. The process has no tendency to have covariances revert to a constant value and should be particularly useful in modeling correlations that jump one or more times. There are many examples of processes that have structural breaks that are unlikely to be reversed. The matrix version of this specification is

$$Q_t = \lambda \varepsilon_{t-1} \varepsilon'_{t-1} + (1 - \lambda)Q_{t-1}. \tag{4.8}$$

Most changes in correlations appear to be temporary and mean reverting. A specification that embodies this assumption is simply analogous to the GARCH$(1, 1)$ process. More precisely, the *mean-reverting* model is an analogue of the scalar diagonal vector GARCH but in terms of the volatility-adjusted data. The specification of the quasi-correlation process between the first two assets is

$$Q_{1,2,t} = \omega_{1,2} + \alpha\varepsilon_{1,t-1}\varepsilon_{2,t-1} + \beta Q_{1,2,t-1}. \tag{4.9}$$

The matrix version of this process is simply

$$Q_t = \Omega + \alpha\varepsilon_{t-1}\varepsilon'_{t-1} + \beta Q_{t-1}. \tag{4.10}$$

This process has two unknown dynamic parameters and $\frac{1}{2}Nx(N-1)$ parameters in the intercept matrix.

Fortunately, a simple estimator for the intercept parameters is available through what is called *correlation targeting*. This will be discussed in more detail below but essentially amounts to using an estimate of the unconditional correlations among the volatility-adjusted random variables. That is, using the estimator

$$\hat{\Omega} = (1 - \alpha - \beta)\bar{R}, \quad \bar{R} \equiv \frac{1}{T}\sum_{t=1}^{T}\varepsilon_t\varepsilon'_t, \tag{4.11}$$

reduces the number of remaining unknown parameters to two. This must be taken into account when evaluating the properties of the estimator, as will be done in chapter 11. The statistical underpinning of this restriction will also be discussed in more detail.

Substituting (4.11) into (4.10) gives the basic form for the mean-reverting DCC model:

$$Q_t = \bar{R} + \alpha(\varepsilon_{t-1}\varepsilon'_{t-1} - \bar{R}) + \beta(Q_{t-1} - \bar{R}). \tag{4.12}$$

The matrix Q is guaranteed to be positive definite if α, β, and $(1 - \alpha - \beta)$ are all positive and if the initial value, Q_1, is positive definite. This is because each subsequent value of Q is simply the sum of positive-definite or positive-semidefinite matrices and therefore must be positive definite.

It is now easy to see how this model behaves. The correlations, which are approximately given by the off-diagonal elements of Q, evolve over time in response to new information on the returns. When returns are moving in the same direction—either they are both moving up or they are both moving down—the correlations will rise above their average level and remain there for a while. Gradually, this information will decay and correlations will fall back to the long-run average. Similarly, when assets move in opposite directions, the correlations will temporarily fall below the unconditional value. The two parameters (α, β) govern the speed of this adjustment. These two parameters will need to be estimated from the data. Notice that this is an extremely parsimonious specification as only two parameters are used regardless of the size of the system being modeled.

A third very useful model is the *asymmetric* DCC, or ADCC, model. This model recognizes that the dynamic adjustment of correlations may be different for negative random variables than it is for positive ones. In equation (4.9) it is clear that correlations will be the same when both variables are positive as when they are both negative. However, for many financial applications it has been observed that correlations increase faster when markets are declining. Thus an additional term is needed in the correlation process:

$$Q_t = \Omega + \alpha \varepsilon_{t-1} \varepsilon'_{t-1} + \gamma \eta_{t-1} \eta'_{t-1} + \beta Q_{t-1}, \quad \eta_t = \min[\varepsilon_t, 0]. \quad (4.13)$$

The product $\eta_{i,t} \eta_{j,t}$ will be nonzero only if both variables are negative. Hence a positive value of γ will give the desired result: that correlations increase more in response to market declines than they do in response to market increases.

Estimation of Ω is a little more complicated in this case than it is in the symmetric case above as it involves an estimate of the unconditional covariance matrix of η as well as of the covariance matrix of ε. The estimation formula is

$$\hat{\Omega} = (1 - \alpha - \beta)\bar{R} - \gamma \bar{N}, \quad \bar{R} \equiv \frac{1}{T} \sum_{t=1}^{T} \varepsilon_t \varepsilon'_t, \quad \bar{N} = \frac{1}{T} \sum_{t=1}^{T} \eta_t \eta'_t. \quad (4.14)$$

In this model it is not so straightforward to find conditions under which Ω will be positive definite. From (4.13) we can see that

$$\left. \begin{array}{c} \varepsilon_t = \eta_t + \pi_t, \quad \pi_t = \max(\varepsilon_t, 0), \\[4pt] \varepsilon_t \varepsilon'_t = \eta_t \eta'_t + \pi_t \pi'_t, \quad \text{since } \eta_t \pi'_t = 0, \\[4pt] \bar{R} = \bar{N} + \bar{P}, \quad \text{where } \bar{P} = \frac{1}{T} \sum_{t=1}^{T} \pi_t \pi'_t. \end{array} \right\} \quad (4.15)$$

Hence (4.14) becomes

$$\Omega = (1 - \alpha - \beta - \gamma)\bar{N} + (1 - \alpha - \beta)\bar{P}. \quad (4.16)$$

Consequently, a sufficient condition for the positive definiteness of $\hat{\Omega}$ is that $(1 - \alpha - \beta - \gamma) > 0$, but a weaker condition has been developed for this case in Engle and Sheppard (2005b).

They pre- and post-multiply (4.14) by the square root of the inverse of the correlation matrix:

$$\bar{R}^{-1/2} \hat{\Omega} \bar{R}^{-1/2} = (1 - \alpha - \beta)I - \gamma [\bar{R}^{-1/2} \bar{N} \bar{R}^{-1/2}]. \quad (4.17)$$

The smallest eigenvalue of this expression will correspond to the largest eigenvalue of the matrix in square brackets. If this largest eigenvalue is δ, then the condition for all eigenvalues to be positive, and therefore for $\hat{\Omega}$ to be positive definite, is

$$1 - \alpha - \beta - \delta\gamma > 0. \tag{4.18}$$

Since δ can be calculated before estimating parameters, this condition is easy to impose and improves the sufficient condition given below (4.16). To see this, write the covariance matrix as the sum of the covariances of the negative part and the covariances of the positive part,

$$\bar{R} = \bar{N} + \bar{P}, \qquad \bar{R}^{-1/2}\bar{N}\bar{R}^{-1/2} = I - \bar{R}^{-1/2}\bar{P}\bar{R}^{-1/2}, \tag{4.19}$$

ensuring that the largest eigenvalue is less than 1.

Many other specifications can be considered for the correlation process. Far more elaborate specifications are used in various papers that allow the process to be different for different variables, that allow shifts in regime, or that introduce factor structures making some components of correlations common across assets. Some of these extensions will be discussed in the next chapter.

4.3 Rescaling in DCC

All of these models specify a process for the matrix Q that delivers a positive-definite quasi-correlation matrix for each time period. However, they do not ensure that this is a correlation matrix. The diagonal elements will be 1 on average but will not be 1 for every observation. To convert these Q processes into correlations, they must be rescaled. The diagonal elements of Q can be calculated by the same formula and used to rescale the Qs to estimate the correlations. The process is simply

$$\rho_{i,j,t} = \frac{Q_{i,j,t}}{\sqrt{Q_{i,i,t}Q_{j,j,t}}}. \tag{4.20}$$

Although the expected value of each of the two denominator terms is 1, they are not estimated to be exactly 1 at all points in time and therefore the Qs may be outside the interval $(-1, 1)$. This equation is called rescaling and its matrix form is given by

$$R_t = \text{diag}\{Q_t\}^{-1/2}Q_t \, \text{diag}\{Q_t\}^{-1/2}. \tag{4.21}$$

Clearly this introduces a nonlinearity into the estimator. Since Q is generally linear in the squares and cross products of the data, this nonlinearity means that R is nonlinear in the squares and cross products and hence cannot be an unbiased estimator of the true correlation. Furthermore, the forecasts are not unbiased. These drawbacks are also inherent in all other multivariate GARCH methods, however, including even simple models such as historical correlations. Intuitively this should not be surprising because correlations are bounded and the data are not.

An important alternative specification to DCC was introduced by Tse and Tsui (2002). They formulate a closely related model:

$$R_t = \bar{R}(1 - \alpha - \beta) + \alpha \hat{r}^k_{t-1} + \beta R_{t-1},$$

$$\hat{r}^k_t = \text{diag}\left[\frac{1}{k}\sum_{s=t-k}^{t}\varepsilon_s\varepsilon'_s\right]^{-1/2}\left[\frac{1}{k}\sum_{s=t-k}^{t}\varepsilon_s\varepsilon'_s\right]\text{diag}\left[\frac{1}{k}\sum_{s=t-k}^{t}\varepsilon_s\varepsilon'_s\right]^{-1/2}.$$

(4.22)

In this expression $\hat{r}^k_t$ is a rolling correlation estimator using the most recent k observations. It will have 1s on the diagonal, and, by recursion, so will all Rs. The matrices will all be positive definite and will all be correlation matrices. The apparent advantage of this approach is that there is no nonlinearity involved. However, it should be clear that the nonlinearity is simply in a different place. It is now in the transformation of the data to get $\hat{r}^k_t$. In this model, the lag structure is determined jointly by the choice of k and (α, β). Because k must be greater than the dimension of the system, that choice is limited. As systems get larger, the ability of this estimator to respond to news is reduced as it can only change correlations by changing $\hat{r}^k_{t-1}$, and when k is large this effect is small. Furthermore, the use of rolling correlations means, of course, that an observation is influential both when it enters the rolling window and when it leaves it.

To examine this effect in several multivariate models, we consider the updating function that gives the correlation for period $t + 1$ in terms of the news in period t and the data prior to this. This is often called a *correlation news impact surface* following Engle and Ng (1993) and Kroner and Ng (1998). Without loss of generality, the correlation between the first two random variables, 1 and 2, can be expressed in terms of the standardized shocks, because the conditional mean and variance are already functions of the prior information:

$$\rho_{t+1} = f(\varepsilon_{1,t}, \varepsilon_{2,t}; \rho_t, \text{parameters}, \text{prior data}),$$

(4.23)

where, as before,

$$\varepsilon_{i,t} = \frac{y_{i,t} - E_{t-1}(y_{i,t})}{\sqrt{V_{t-1}(y_{i,t})}}.$$

(4.24)

For example, the exponential smoother has a covariance matrix given by

$$H_{t+1} = \lambda H_t + (1 - \lambda)y_t y'_t,$$

(4.25)

assuming that y is a vector of zero-mean random variables. The correlation between the first two random variables is given by

$$
\begin{aligned}
\rho_{t+1} &= \frac{H_{12,t+1}}{\sqrt{H_{11,t+1}H_{22,t+1}}} \\
&= \frac{\lambda H_{12,t} + (1-\lambda)y_{1,t}y_{2,t}}{\sqrt{(\lambda H_{11,t} + (1-\lambda)y_{1,t}^2)(\lambda H_{22,t} + (1-\lambda)y_{2,t}^2)}} \\
&= \frac{\lambda \rho_t + (1-\lambda)\varepsilon_{1,t}\varepsilon_{2,t}}{\sqrt{(\lambda + (1-\lambda)\varepsilon_{1,t}^2)(\lambda + (1-\lambda)\varepsilon_{2,t}^2)}}.
\end{aligned}
\tag{4.26}
$$

This expression ranges from ρ_t when $\lambda = 1$ to ± 1 when $\lambda = 0$. It is a weighted average of the previous estimated correlation and the correlation estimated from the most recent observation, which is simply $+1$ or -1. When either of the shocks are close to 0, the correlation is unchanged. When they are both large, the correlation approaches $+1$ or -1. The single parameter λ regulates this transition.

A plot of the correlation news impact surface of equation (4.26) for various values of ρ_t would be three dimensional. However, by examining the cases when the two shocks are equal or when they are of opposite sign, the main features of the two-dimensional surface can be presented more easily. In the following plots, the correlation news impact curve is plotted for $\varepsilon_{1,t} = \varepsilon_{2,t}$ and for $\varepsilon_{1,t} = -\varepsilon_{2,t}$. These curves are drawn for values of $\rho_t = (0, 0.5, 0.9)$, and $\lambda = 0.94$ as utilized by RiskMetrics.

The correlation news impact surface in figure 4.1 shows the predicted correlation for the exponential smoother if both returns have shocks that are the same number of standard deviations. When epsilon is 0, the correlation remains at its previous value of 0, 0.5, or 0.9. The upward-sloping curves correspond to shocks that have the same sign and the downward-sloping curves correspond to shocks that are of opposite sign. When the previous correlation is 0.9, positively related shocks increase the correlation slightly but only gradually does it approach 1. Negatively related shocks, however, have a very substantial effect in reducing the correlations, ultimately to -1. Of course if the correlation was previously 0.9, then large shocks of opposite sign are very surprising and informative. When the current estimate is 0, the upward-sloping and downward-sloping curves are symmetrical. The clear message to take from this picture is that the curves asymptote to $+1$ or -1, but that this occurs only for very large shocks. The x-axis of figure 4.1 is on a scale of 0 to 8 standard deviations. Hence the right segment of the graph will only be relevant for very extreme events. When the shocks are small, the curve is approximately quadratic with a sign that depends upon the sign of the

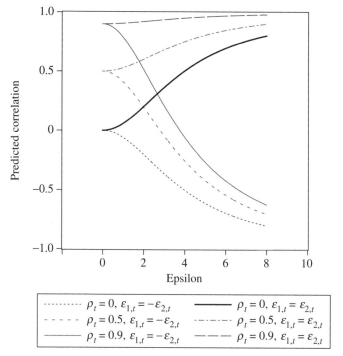

Figure 4.1. Correlation news impact curves: exponential smoother. Predicted correlation is a function of this period's correlation and epsilon, where epsilon is a return measured in standard deviations.

cross product of shocks. This curvature continues until about 1.5 standard deviations. The curve is then linear in the shocks out to perhaps 3 standard deviations and then it rebends as it approaches the asymptote. The curves for negative values of epsilon are mirror images of these as this is a symmetric model.

A similar result can be found for the historical, or rolling, correlation estimator. For an estimator with an expanding window, the covariance matrix can be written as

$$H_{t+1} = \frac{N-1}{N} H_t + \frac{1}{N} \varepsilon_t \varepsilon_t', \tag{4.27}$$

which is the same as (4.25) when $\lambda = 1 - 1/N$. The parameter value $\lambda = 0.94$ corresponds approximately to seventeen period historical correlations. When longer windows are used, the correlation news impact curves are smoother so that larger values of epsilon are needed to change correlations. For a fixed-window historical correlation, there is also a shock from the observation N periods ago that is dropped. This is basically another term that enters into the calculation and adds a third element to the updating relation. This is an undesirable feature of rolling correlations.

The correlation news impact surface can be compared for various estimators. For the DCC model described above, the correlation between the first two assets can be expressed as

$$
\begin{aligned}
\rho_{t+1} &= \frac{Q_{12,t+1}}{\sqrt{Q_{11,t+1}Q_{22,t+1}}} \\
&= \frac{\bar{R}_{12}(1 - \alpha - \beta) + \beta Q_{12,t} + \alpha \varepsilon_{1,t}\varepsilon_{2,t}}{\sqrt{(1 - \alpha - \beta + \beta Q_{11,t} + \alpha \varepsilon_{1,t}^2)(1 - \alpha - \beta + \beta Q_{22,t} + \alpha \varepsilon_{2,t}^2)}}.
\end{aligned}
\tag{4.28}
$$

This expression can be seen to be a weighted average of three correlation estimates: the unconditional correlation $\bar{R}_{12}$, the previous correlation estimate $Q_{12,t}/\sqrt{Q_{11,t}Q_{22,t}}$, and $+1$ or -1. At each point in time, there is a correlation forecast that would result if the shock were 0. This forecast reflects a small but important amount of mean reversion. If the unconditional correlation is set to the same values of $(0, 0.5, 0.9)$ as given above for the exponential smoother, then the only difference will be dependent on the size of α compared with $1 - \lambda$. Typically for DCC, the estimates of α are only 0.01, hence the curves will be flatter than for the exponential smoother with the RiskMetrics value of 0.06. For DCC the correlation news impact curve is shown in figure 4.2 for typical values of $(\alpha, \beta) = (0.01, 0.97)$. These curves appear to be quadratic for about 3 standard deviations; for large shocks, even up to 8 standard deviations, the response is still nearly linear. The point of inflection is at 6 standard deviations.

It is interesting to apply the same analysis to the model of Tse and Tsui (2002). Their model estimates the correlation directly without the rescaling called for by DCC. The prediction for correlations is given by

$$
R_t = \bar{R}(1 - \alpha - \beta) + \alpha \hat{r}_{t-1}^k + \beta R_{t-1}, \tag{4.29}
$$

$$
\hat{r}_t^k = \mathrm{diag}\left[\frac{1}{k}\sum_{s=t-k}^{t}\varepsilon_s\varepsilon_s'\right]^{-1/2}\left[\frac{1}{k}\sum_{s=t-k}^{t}\varepsilon_s\varepsilon_s'\right]\mathrm{diag}\left[\frac{1}{k}\sum_{s=t-k}^{t}\varepsilon_s\varepsilon_s'\right]^{-1/2}.
\tag{4.30}
$$

In this model, no rescaling is needed; however, the rolling correlation estimator in (4.30) is itself a nonlinear function of the data. The correlation news impact surface can be constructed for this model as well. The

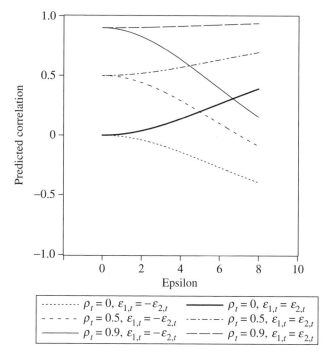

Figure 4.2. Correlation news impact curves: DCC. Predicted correlation is a function of this period correlations and epsilons, where epsilon is a return measured in standard deviations.

predicted correlation between the first two returns is given by

$$\rho_{t+1} = \bar{R}_{12}(1 - \alpha - \beta) + \beta\rho_t$$
$$+ \alpha \frac{\sum_{j=1}^{k} \varepsilon_{1,t-j}\varepsilon_{2,t-j} + \varepsilon_{1,t}\varepsilon_{2,t}}{\sqrt{(\sum_{j=1}^{k} \varepsilon_{1,t-j}^2 + \varepsilon_{1,t}^2)(\sum_{j=1}^{k} \varepsilon_{2,t-j}^2 + \varepsilon_{2,t}^2)}}. \quad (4.31)$$

In their empirical examples, Tse and Tsui chose $k = n$, the number of variables being correlated. This was either 2 or 3 in their examples. Since the last term cannot exceed $+1$ or -1, the maximum impact of the updating is $+\alpha$ or $-\alpha$. Thus the forecast correlation can never change very much in response to one observation. Moreover, when k is small, it reaches this value for modest shocks. In figure 4.2 the quadratic segment is very short. The point of inflection is at 1 standard deviation and the impact of larger shocks declines, so any shock of more than 4 standard deviations will have roughly the same effect. In fact, the forecast correlation can never approach the limit of unity even if there is an infinite sequence of perfectly correlated shocks. The correlation news impact

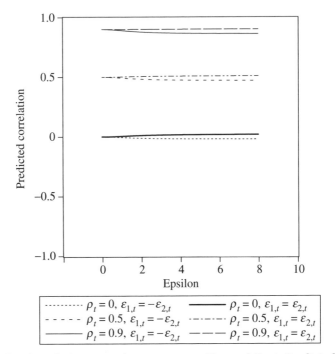

Figure 4.3. Correlation news impact curves: Tse and Tsui. Predicted correlation is a function of this period correlations and epsilons, where epsilon is a return measured in standard deviations.

surface corresponding to this model is given in figure 4.3. The coefficient α ranged from 0.01 to 0.03 in Tse and Tsui's empirical examples and here a value of 0.02 is used with $k = 3$.

Figure 4.4 shows the correlation impact curves for an asymmetric version of DCC. In this model the parameters are taken to be $(\alpha, \beta, \gamma) = (0.018, 0.97, 0.014)$, as estimated for a pair of large-capitalization equities. In this case the curves are shown for $\varepsilon_1 = \varepsilon_2$ when both are up and when both are down. Notice that the correlations increase when both are down and especially when the shocks are very large.

Correlation news impact surfaces can be constructed for the wide range of multivariate GARCH models discussed in chapter 3. These will all have a similar structure but may have quite different magnitudes. For all the models that are linear in squares and cross products, such as the vec and BEKK models, a single sufficiently extreme observation will drive the forecast correlation to 1. In models that are not described as "diagonal," there will also be effects from shocks to other variables. The response of models with asymmetries will be different for negative values of epsilon than for positive values of epsilon.

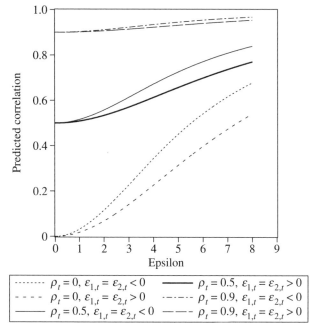

Figure 4.4. Correlation news impact curves: asymmetric DCC. Predicted correlation is a function of this period correlations and epsilons, where epsilon is a return measured in standard deviations.

4.4 Estimation of the DCC Model

Estimation of the DCC model can be formulated as a maximum-likelihood problem once a specific distributional assumption is made about the data. It will be assumed that the data are multivariate normal with the given mean and covariance structure. Fortunately, the estimator is a quasi-maximum-likelihood estimator (QMLE), in the sense that it will be consistent but inefficient, if the mean and covariance models are correctly specified even if other distributional assumptions are incorrect. This is a consequence of the theorem for multivariate GARCH processes in Bollerslev and Wooldridge (1992). Full maximum-likelihood estimation of the non-Gaussian version of these models has been proposed in several papers.

Just as Bollerslev (1987) introduces a t distribution for univariate returns, Bauwens and Laurent (2005) use a multivariate skew distribution for multivariate GARCH models, Cajigas and Urga (2006) propose a Laplace distribution, and Pelagatti (2004) uses elliptical distributions. An attractive class of non-Gaussian joint distributions based on Gaussian copulas or Student t copulas was described in chapter 2 using meta-distributions and appears to provide a simple computational approach to such estimation problems.

To be precise about the model being considered, it will be fully restated here.

Assume

$$
\left.\begin{aligned}
y_t | \mathbb{F}_{t-1} &\sim N(0, D_t R_t D_t), \\
D_t^2 &= \operatorname{diag}\{H_t\}, \quad H_t = V_{t-1}(y_t), \\
H_{i,i,t} &= \omega_i + \alpha_i y_{i,t-1}^2 + \beta_i H_{i,i,t-1}, \\
\varepsilon_t &= D_t^{-1} y_t, \\
R_t &= \operatorname{diag}\{Q_t\}^{-1/2} Q_t \operatorname{diag}\{Q_t\}^{-1/2}, \\
Q_t &= \Omega + \alpha \varepsilon_{t-1} \varepsilon_{t-1}' + \beta Q_{t-1},
\end{aligned}\right\}
\tag{4.32}
$$

where, α, β, and $\{\alpha_i, \beta_i\}$ for all i are nonnegative with a sum less than unity.

The log likelihood for a data set $\{y_1, \ldots, y_T\}$ can then be written

$$
\begin{aligned}
L &= -\frac{1}{2} \sum_t \left(n \log(2\pi) + \log|H_t| + y_t' H_t^{-1} y_t \right) \\
&= -\frac{1}{2} \sum_t \left(n \log(2\pi) + \log|D_t R_t D_t| + y_t' D_t^{-1} R_t^{-1} D_t^{-1} y_t \right) \\
&= -\frac{1}{2} \sum_t \left(n \log(2\pi) + 2\log|D_t| + \log|R_t| + \varepsilon_t' R_t^{-1} \varepsilon_t \right) \\
&= -\frac{1}{2} \sum_t \left(n \log(2\pi) + 2\log|D_t| + y_t' D_t^2 y_t - \varepsilon_t' \varepsilon_t + \log|R_t| + \varepsilon_t' R_t^{-1} \varepsilon_t \right).
\end{aligned}
\tag{4.33}
$$

This log likelihood function has a standard form and can simply be maximized with respect to all the parameters in the model. These are the parameters in the variances and the parameters in the correlation process.

However, as can be seen from the last line of (4.33), the log likelihood can be divided into two parts. The first three terms simply contain the variance parameters and the data. The last three terms contain the correlation parameters and the volatility-adjusted data:

$$
\begin{aligned}
L(\{y_1, &\ldots, y_T\}; \text{all parameters}) \\
&= L_1(\{y\}; \text{variance parameters}) + L_2(\{\varepsilon\}; \text{correlation parameters}).
\end{aligned}
\tag{4.34}
$$

A two-step estimation method can naturally be applied here. The first step is to maximize the variance part of the likelihood function. The solution in this case is simply to compute univariate GARCH models for each of the series. These are treated as independent in this calculation. The second step is to take the standardized residuals from the first step and maximize the second log likelihood function with respect to the correlation parameters. This step is clearly conditional on the estimated variance parameters. This two-step estimator is not equivalent to full maximum likelihood but is much simpler and is consistent. In fact, it is very close to full maximum likelihood in most cases. The fourth line of (4.33) shows how the two sections can be separated into well-defined likelihood functions which can each be maximized. The term in the sum of squared epsilons can be ignored as it does not depend upon the parameters being optimized.

The two-step version of the CCC model is particularly simple to estimate as the conditional correlation matrix is constant and the maximum-likelihood estimator is just the sample covariance matrix:

$$\bar{R} = \frac{1}{T} \sum \varepsilon_t \varepsilon_t'. \tag{4.35}$$

If the variances are not exactly unity, then the correlation matrix may be more appropriate and is generally used. Bollerslev proposes a joint estimation of volatility and correlation but the two-step method is much simpler and will be consistent. More details on maximum-likelihood estimation and inference are presented later.

The two-step estimation of the mean-reverting DCC equation in (4.9) is similarly accomplished by first estimating GARCH equations for each asset and then maximizing the log likelihood of the DCC model using the standardized residuals as data. The specification of the correlation matrix is

$$Q_t = \Omega + \alpha \varepsilon_{t-1} \varepsilon_{t-1}' + \beta Q_{t-1}. \tag{4.36}$$

When this is estimated by correlation targeting this becomes a *three-step* estimator. First, estimate univariate GARCH models. Second, compute the sample correlation matrix of the standardized residuals. Third, impose the restriction

$$\Omega = \bar{R}(1 - \alpha - \beta) \tag{4.37}$$

or

$$\left.\begin{aligned} Q_t &= \bar{R}(1 - \alpha - \beta) + \alpha \varepsilon_t \varepsilon_t' + \beta Q_{t-1}, \\ R_t &= \mathrm{diag}\{Q_t\}^{-1/2} Q_t \, \mathrm{diag}\{Q_t\}^{-1/2}, \end{aligned}\right\} \tag{4.38}$$

and maximize the log likelihood function:

$$L_2 = -\frac{1}{2} \sum_{t=1}^{T} [\log |R_t| + \varepsilon_t' R_t^{-1} \varepsilon_t]. \tag{4.39}$$

The model in (4.36) has $2 + \frac{1}{2}n(n-1)$ parameters, while the model in (4.38) has only 2 regardless of the number of variables in the system. This great economy in parameters arises from the use of the auxiliary estimator of the correlations. The use of a moments estimator for the unconditional correlations combined with maximum likelihood for the other parameters is called *correlation targeting* and is directly analogous to the *variance-targeting* estimators often used for variance models. The statistical properties of such an estimator will be analyzed in chapter 11 along with the two-step/three-step estimation method. In the next chapter, the two-step/three-step maximum-likelihood estimator of the integrated and mean-reverting models will be examined empirically and from a Monte Carlo point of view.

Various diagnostic tests have been developed for DCC models. For some examples see Engle and Sheppard (2005a), Eklund and Terasvirta (2007), and He and Terasvirta (2004). The most satisfactory approach to inference is based on the GMM formulation of the two-step/three-step nature of the DCC model, as discussed in chapter 11. There remains a need for better diagnostic tests for this class of models.

DCC Performance

5.1 Monte Carlo Performance of DCC

Our first task is to discuss a Monte Carlo experiment reported in Engle (2002a). In this case the true correlation structure is known. Several estimators can compete to approximate it. The data generating processes are deterministic time series processes that can be approximated by the conditional covariance estimators. The simplest assumptions are that the correlations are constant or that they change only once during the sample period. More complex assumptions allow the correlations to follow sine waves or ramps so that they range from zero to one. These are mean-reverting specifications but they differ in the best approach to adapting the past information. Pictures of these processes are given in figure 5.1.

These correlations follow the mathematical processes:

CONSTANT, $\rho_t = 0.9$;

SINE, $\rho_t = 0.5 + 0.4 \cos(2\pi t/200)$;

FAST SINE, $\rho_t = 0.5 + 0.4 \cos(2\pi t/20)$;

STEP, $\rho_t = 0.9 - 0.5(t > 500)$;

RAMP, $\rho_t = \mathrm{mod}(t/200)$.

To complete the specification, the volatilities and correlations are given by

$$\left. \begin{aligned} h_{1,t} &= 0.01 + .05r_{1,t-1}^2 + 0.94h_{1,t-1}, \\ h_{2,t} &= 0.5 + 0.2r_{2,t-1}^2 + 0.5h_{2,t-1}, \\ \begin{pmatrix} \varepsilon_{1,t} \\ \varepsilon_{2,t} \end{pmatrix} &\sim N\left[0, \begin{pmatrix} 1 & \rho_t \\ \rho_t & 1 \end{pmatrix}\right], \\ r_{1,t} &= \sqrt{h_{1,t}}\varepsilon_{1,t}, r_{2,t} = \sqrt{h_{2,t}}\varepsilon_{2,t}. \end{aligned} \right\} \qquad (5.1)$$

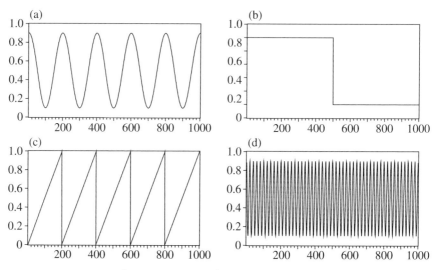

Figure 5.1. Correlation experiments:
(a) SINE, (b) STEP, (c) RAMP, (d) FAST SINE.

In one experiment the epsilons are Student t with four degrees of freedom. One series is highly persistent while the other is not. Each replication simulates one thousand observations, which corresponds to approximately four years of daily data. This is a moderate time series from a daily data point of view. Two hundred replications are constructed for each scenario.

Seven different methods are used to estimate the correlations: two multivariate GARCH models, orthogonal GARCH, the integrated DCC model and the mean-reverting DCC model, the exponential smoother from RiskMetrics, and the familiar 100-day moving-average model. The multivariate GARCH models were described in more detail in chapter 3. The orthogonal GARCH method is implemented by a pair of GARCH regressions:

$$\left. \begin{array}{c} y_{2,t} = \beta y_{1,t} + \sqrt{h_{2/1,t}}\, \varepsilon_{2,t}, \quad h_{2/1,t} \sim \text{GARCH}(1,1), \\[2mm] y_{1,t} = \sqrt{h_{1,t}}\, \varepsilon_{1,t}, \quad h_{1,t} \sim \text{GARCH}(1,1), \\[2mm] \rho_t = \dfrac{\beta h_{1,t}}{\sqrt{(\beta^2 h_{1,t} + h_{2/1,t})(h_{1,t})}}. \end{array} \right\} \quad (5.2)$$

The methods and their descriptions are as follows.

Scal BEKK: a scalar version of (3.11) with variance targeting.

Diag BEKK: a diagonal version of (3.9) with variance targeting.

DCC INT: dynamic conditional correlation for an integrated process.

DCC MR: dynamic conditional correlation with a mean-reverting model.

MA 100: a moving average of 100 days.

OGARCH: orthogonal GARCH, as in (5.2).

EX0.06: exponential smoothing with a parameter of 0.06.

Table 5.1 presents the results for the mean absolute error (MAE) for the seven estimators for six experiments with 200 replications. In four of the six cases the DCC mean-reverting model has the smallest MAE. When these errors are summed over all cases, this model is the best one. Very close second- and third-place models are the integrated DCC and the scalar BEKK. Interestingly, when the conditional correlation is constant, the mean-reverting model is not as good as the integrated model, and when there is a step in the correlation, the best model is the moving average. The MAEs in correlation estimation are very small for several of the estimators in the constancy case, and are largest for all estimators for the fast sine experiment.

The plot in figure 5.2 shows the sum of the MAEs for each of the experiments. The height is therefore a measure of the average performance of these tasks. The best method is clearly the mean-reverting DCC followed by the integrated DCC. In Engle (2002a) a variety of other diagnostic criteria are applied to these estimators with the finding that the DCC methods provide very good performance for both real and simulated data.

5.2 Empirical Performance

The methods described above provide a powerful set of tools for forecasting correlations between financial assets. There is now a very extensive literature applying the DCC model to a wide range of assets as well as nonfinancial data sets.

Bautista (2006) looks at exchange rate correlations and how they depend upon currency regimes. Ciner (2006) looks at linkages between North American Free Trade Agreement (NAFTA) equity markets. Crespo Cuaresma and Wojcik (2006) look at correlations between new European Union (EU) member countries. Bodurtha and Mark (1991) look at the role of time variation in risks and returns in tests of the Capital Asset Pricing Model (CAPM). Chan et al. (2005b) and Chan et al. (2005a) look at patents and tourism. Chandra (2005) examines extremes in correlation and Chandra (2006) looks at deterministic effects such as the day of the week. Chiang et al. (2007) look at Asian currency correlations. Ang and Chen (2002) show the importance of asymmetry in equity portfolios. Maghrebi et al. (2006) focus on asymmetries in the covariance

Table 5.1. Mean absolute error of correlation estimates.

Model	Scal BEKK	Diag BEKK	DCC MR	DCC INT	MA 100	OGARCH	EX0.06
FAST SINE	0.2292	0.2307	**0.2260**	0.2555	0.2599	0.2474	0.2737
SINE	0.1422	0.1451	**0.1381**	0.1455	0.3038	0.2245	0.1541
STEP	0.0859	0.0931	0.0709	0.0686	**0.0652**	0.1566	0.081
RAMP	0.1610	0.1631	**0.1546**	0.1596	0.2828	0.2277	0.1601
CONSTANT	0.0273	0.0276	0.0070	**0.0067**	0.0185	0.0449	0.0276
T(4) SINE	0.1595	0.1668	**0.1478**	0.1583	0.3016	0.2423	0.1599

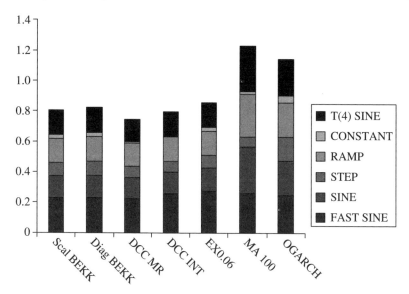

Figure 5.2. Sum of MAE correlation estimates.

processes of Asian currency and equity markets. Milunovich and Thorp (2006) demonstrate the importance of volatility spillovers in portfolio formation with European equities. Duffee (2005) examines the correlation between equities and consumption. Engle and Sheppard (2005a) examine equity correlations of industry portfolios. Flavin (2006) uses multivariate GARCH to evaluate the risk aversion of fund managers. Goetzmann et al. (2005) evaluate correlations from a long-run economic perspective. Guirguis and Vogel (2006) look at the correlations between real-estate prices and find an asymmetric model. Lee (2006) uses the DCC to examine comovement of inflation and output. Yang et al. (2006) examine the dynamic correlations of equity and industry indices and find that the industry correlations are mostly lower than national correlations. Bystrom (2004) uses the orthogonal GARCH model in a multivariate analysis of Nordic equity markets. Kearney and Patton (2000) and Kearney and Poti (2004) model European equity and currency markets. Koutmos (1999) models the asymmetry in G7 equity returns. Ku et al. (2007) and Kuper and Lestano (2007) model optimal hedge ratios in European and Asian markets. Kim et al. (2006) model EU stock and bond returns, examining the impact of integration.

In this chapter several different correlation models will be applied to pairs of financial returns. These will allow us to compare different methods using a common data set. The first example is the correlation between returns on stocks and bonds. This example is of great

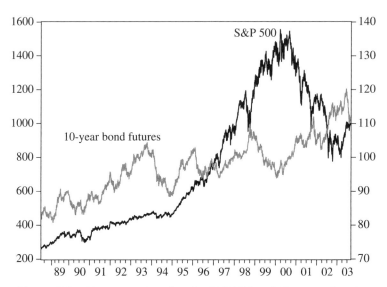

Figure 5.3. Futures prices for the S&P 500 and Treasury bonds.

importance in asset allocation, where an investor must choose the proportions of his assets to hold in equities and in fixed income. The optimal choice depends on the volatilities and correlations between these asset classes as well as on expectations of the returns on each asset.

Engle and Colacito (2006) examine the daily returns on futures prices on the S&P 500 and on 10 Year Treasury bonds. These data are continuously compounded returns from Datastream and are based on the near-term contract, which is rolled over in the month of maturity. The data from August 26, 1988, through August 28, 2003, are plotted in figure 5.3.

The prolonged surge in equity prices to 2000 is highly evident in this figure, as is the subsequent decline as the tech bubble burst. At the end of the sample, the beginning of a rebound can be seen. The bond price increases correspond to the gradual decline in long-term interest rates over the decade and a half. There are, however, several episodes of rising rates as the Federal Reserve (Fed) sought to quell the equity bubble in 1994–95 and in 1999 and 2000.

The correlation between stock and bond returns is close to zero on average but it is not at all constant. Looking at historical correlations with one-month (22 days) and one-year (250 days) rolling windows, we see in figure 5.4 the changes in correlations. These are generally positive until approximately 1998 and thereafter they are negative.

Most asset returns are positively correlated, since good news for the economy increases the value of most assets. This is not always true of

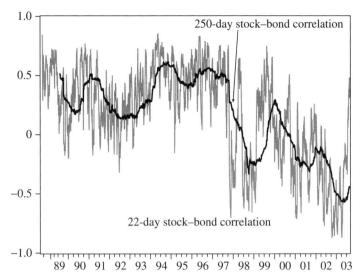

Figure 5.4. Historical stock-bond correlations.

bonds, particularly when the economy is close to full employment. During these periods, good news for the economy is bad news for bonds as it signals higher inflation in the future or influences Fed policy toward raising interest rates; conversely, bad news for the economy becomes good news for the bond market.

The basic economic explanation for time variation in stock and bond correlations is given in Campbell and Ammer (1993). Anderssona et al. (2008) and de Goeij and Marquering (2004) investigate empirically the macroeconomic determinants of stock-bond correlations.

Looking at figure 5.4, we see that as the economy overheated in the late 1990s and Alan Greenspan complained about "irrational exuberance," the correlations between the stock and bond markets passed into the negative region. This was exaggerated by the LTCM crisis and the Russian bond default, both of which lowered bond prices in a time of rising equity markets. The negative correlations continued through the market declines of 2001-2 as the Fed lowered rates, stimulating the bond market but not the equity market.

Only at the very end of the sample period is there evidence that the correlations are beginning to return toward positive values.

The two correlation curves in figure 5.4 look quite different. The annual historical correlations are much smoother and easier to interpret. The monthly correlations have a great deal of volatility, which looks like noise. However, the annual correlations can also be seen to lag behind

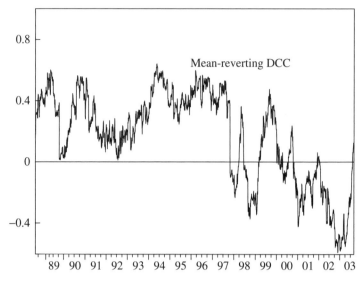

Figure 5.5. Stock–bond correlations.

the monthly correlations and miss some of the features. There is no satisfactory statistical criterion for choosing between these measures.

These estimates can be compared with estimates based on a statistical model such as multivariate GARCH. In figure 5.5 the conditional correlations from mean-reverting DCC, as described in (4.12), are exhibited. As can be seen, these parameters show the same general pattern as the historical correlations; however, they are less volatile than the 22-day correlations and yet pick up the shape better than the one-year correlations. For example, they return all the way to 0 at the end of the sample and show drops in both 1997 and 1998. These features are clear in the monthly correlations. These correlations come from estimated parameters ($\{\alpha = 0.023,\ \beta = 0.973\}$). Notice that the sum is very close to 1, giving rise to substantial persistence of deviations from the unconditional level.

These estimates can be compared with a variety of other estimators. The fit of the asymmetric DCC model to this data set shows no evidence of asymmetric correlations. In fact, this is not generally a feature of bond returns as much as it is of equity returns. The estimated parameter for the integrated DCC model is $\lambda = 0.022$, which is quite close to the estimate of α for the mean-reverting model, leading to correlations that are quite close to those for the mean-reverting model.

In fact, for this data set, the estimates from several multivariate GARCH models also look quite similar. These are shown in figure 5.6. The figure includes plots of the scalar MGARCH defined in (3.11), the

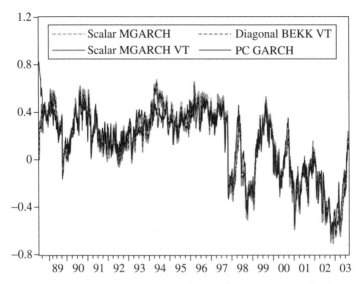

Figure 5.6. Stock–bond correlations by various methods.

scalar MGARCH with variance targeting, and the diagonal BEKK model defined in (3.10). Finally, the principal-component version of OGARCH is presented as defined in (3.34). It is difficult to see the differences between these curves.

Engle and Colacito (2006) consider the asset allocation implications of using these different correlation estimates. Although the portfolios differ over time, the differences in performance between the top models are not statistically significant. The best models are the mean-reverting DCC, the asymmetric DCC, and the diagonal BEKK with variance targeting. It is not possible to reject the hypothesis that these are equivalent for this data set.

Another interesting data set that reveals different features of the methods is time series data on U.S. large-capitalization equities. The American Express Company (AXP) and General Electric (GE) are both businesses that have evolved over the last decade. Although they were originally a travel company and an appliance manufacturer, respectively, they have each added important financial service operations. See figure 5.7, which shows the growth in revenue at GE due to commercial and consumer finance.

As these companies' lines of business change, it is natural for the correlations to change. This is one of the key reasons for changing correlations over long periods of time. In this case the correlations between GE and AXP naturally increase. The correlations as estimated by integrated

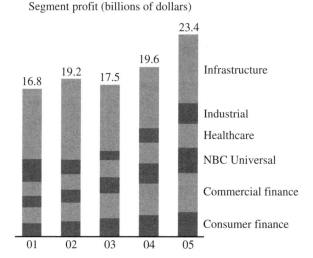

Figure 5.7. General Electric profit segments.

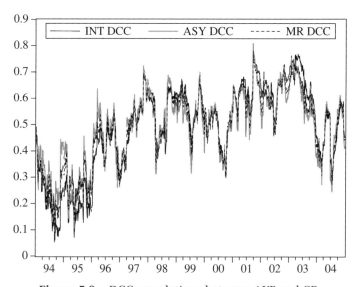

Figure 5.8. DCC correlations between AXP and GE.

DCC (DCC INT), mean-reverting DCC (DCC MR), and asymmetric DCC (DCC ASY) are presented in figure 5.8.

The rise in correlations is unmistakable. Correlations averaged about 0.2 in 1994 and 0.7 in 2002. However, it is clear that this process has been influenced by more than just the simple change in business plan. The increase in volatility of the market in 1997–2002 will contribute to increasing correlations in simple one-factor models such as the CAPM.

Table 5.2. DCC parameters for AXP and GE.

Method	α	T-statistic	β	T-statistic	γ	T-statistic
DCC INT	0.017	7.4	0.983			
DCC MR	0.019	6.4	0.976	219.5		
DCC ASY	0.018	3.3	0.967	161.1	0.015	1.98

Changing interest rate policy will also add to the correlations of interest-rate-sensitive businesses. Both of these effects can be seen by visual examination of these plots.

There are some differences between these three models for this data set. The DCC ASY model has the highest and the lowest correlations; this estimate has the most volatility. The DCC INT has the highest at the end of the sample and the lowest near the beginning. That is, the integrated model has the highest correlations when they are in a high state and the lowest correlations when they are in a low state. The estimated parameter values are given in table 5.2.

The coefficient, α, is very similar for all three models. In the DCC INT this is interpreted as λ, and β would be interpreted as $1 - \lambda$. The sum of α and β is above 0.99 for the DCC MR model, indicating substantial persistence of shocks. In the DCC ASY model, the coefficient for joint negative returns is barely significant but is almost the same size as the coefficient α. Thus, observations where both stocks are falling will be associated with almost twice as big an increase in correlation as those when both are rising. The sum of the three coefficients in the asymmetric model slightly exceeds 1 and thus does not quite satisfy the simplest sufficient condition for covariance stationarity. To show that this model is covariance stationary, condition (4.18) can be checked.

A second interesting example is the correlation between Boeing (BA) and General Motors (GM) shown in figure 5.9. These are two companies in different industries: aircraft and cars. As such, the correlations might not be expected to be particularly large. On average this correlation is 0.27— similar to the correlation between random large-capitalization stocks. After growing steadily through the 1990s, the two stocks decoupled in 1999 and 2000. When one went up, the other went down in the frenzy over the Internet market. However, the correlations rose dramatically after 9/11, presumably because both industries are very sensitive to energy prices. Energy prices began to rise dramatically some time later, but the correlations rose immediately based on the news.

The interpretation of this event is very important in understanding why correlations change over time. A factor that is relatively stable

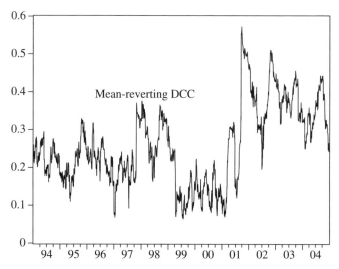

Figure 5.9. Correlations of Boeing and General Motors.

suddenly becomes unstable. As a result, all companies that are sensitive to this factor have more volatile and more correlated returns. This could be interpreted as a missing factor in a factor model; however, there will always be missing factors. Hence it is not possible to anticipate such a factor; it is better to measure correlations dynamically so that these factors do not impair our financial decision making.

We now apply the mean-reverting DCC model to a selected set of large-capitalization stocks over the same sample period. In this data set, the model is fitted separately for each pair of stocks and then the estimates are combined into a system using the MacGyver method discussed in chapter 6.

The stocks are AXP, BA, Disney (DIS), GE, GM, and JPMorgan (JPM). They represent a variety of segments of the U.S. economy. The results are shown in figure 5.10. For comparison, figure 5.11 shows the 100-day historical correlations for the same set of stocks.

The patterns of correlations clearly differ across stocks. Some are basically rising over the decade, while others fall sharply during the Internet bubble. For example, the financial stocks AXP, JPM, and GE showed increasing correlations throughout, reaching almost 0.8 in 2003. On the other hand, the correlations of Boeing and Disney with General Electric, American Express and JPMorgan rose until about 1997 and then fell, becoming 0 or slightly negative. These stocks are essentially uncorrelated with the other stocks and with each other during the Internet period.

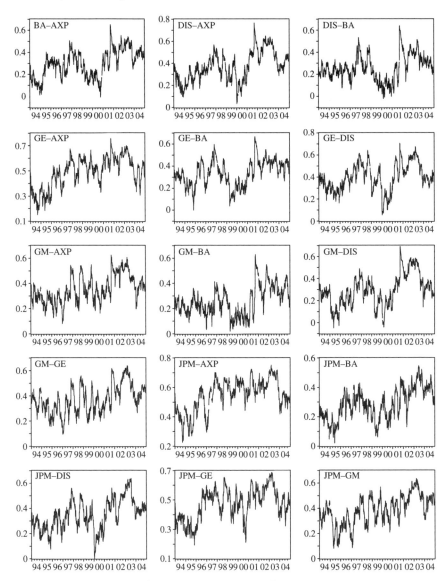

Figure 5.10. DCC correlations.

These patterns of correlation movement appear to be mean reverting, in the sense that movements are eventually reversed. This is simply an observation and does not mean that other movements in correlations would not be permanent shifts.

Finally, these estimates can be used to construct conditional betas. The conditional beta of a stock is the conditional covariance with the market index divided by the conditional variance of the market index.

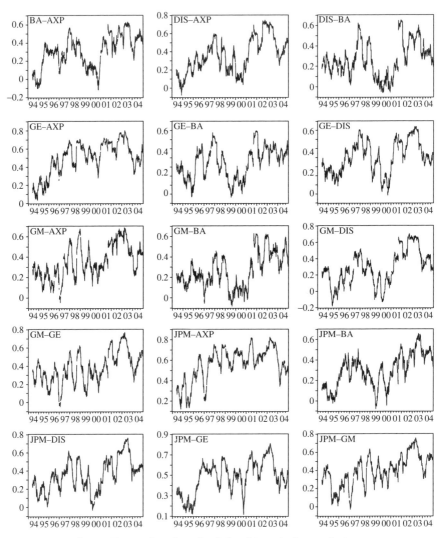

Figure 5.11. One-hundred-day historical correlations.

This is easily calculated for stock i and market return S&P 500, as

$$\beta_{i,t} = \frac{E_{t-1}(r_{i,t}r_{SP,t})}{V_{t-1}(r_{SP,t})} = \rho_{i,SP,t}\sqrt{h_{i,t}/h_{SP,t}}. \quad (5.3)$$

The results for the mean-reverting DCC estimates of conditional correlations are given in figure 5.12. As can be seen, these show substantial fluctuation over time. The estimates fluctuate around 1 but are often below 1. Boeing and Disney show substantial reductions in beta in 1998–2000, while the other stocks show a small but not persistent reduction in beta. The beta for General Motors declines from 1995 onward and never really

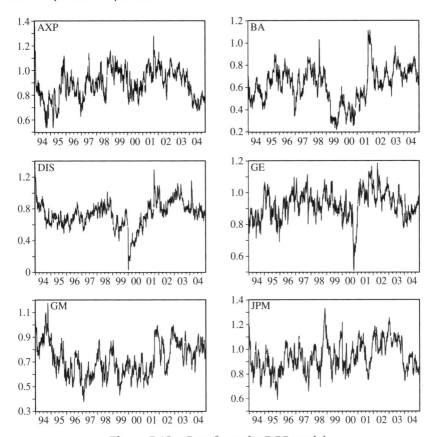

Figure 5.12. Beta from the DCC model.

recovers. At the end of the sample, several of the betas decline as market volatilities decline, while others rise.

These estimates use the most recent data to calculate an estimate of the beta that will be relevant for the next period. If this beta is rapidly mean reverting, then it may not have relevance for asset pricing, but it would still have relevance for risk assessment and for hedging. If, however, there are structural changes in the betas or mean reversion happens only slowly, then there could be important asset pricing implications.

This chapter has developed the basic models for dynamic conditional correlation. The empirical methods to implement these models have been demonstrated in both Monte Carlo and real data contexts. There is much promise in the quality of the results and the simplicity of the method. It remains to show how these methods can improve financial decisions and how to extend these methods to richer specifications and larger data environments.

6
The MacGyver Method

The problem of estimating correlation matrices for large systems might appear to have been solved in previous chapters. However, there are three reasons to believe that we do not yet have a full solution. First, the evaluation of the log likelihood function requires inversion of matrices, R_t, which are full $n \times n$ matrices, for each observation. To maximize the likelihood function, it is necessary to evaluate the log likelihood for many parameter values and consequently invert a great many $n \times n$ matrices. Convergence is not guaranteed and sometimes it fails or is sensitive to starting values. These numerical problems can surely be alleviated but ultimately, for very large n, the numerical issues will dominate. Second, Engle and Sheppard (2005b) show that in correctly specified models with simulated data there is a downward bias in α when n is large and T is only somewhat bigger than n. Thus the correlations are estimated to be smoother and less variable when a large number of assets are considered than when a small number of assets are considered. Third, there may be structure in correlations that is not incorporated in this specification. This of course depends on the economics of the data in question but the introduction of the FACTOR DCC model in chapter 8 is a response to this issue.

In this section I will introduce a new estimation method that is designed to solve the first two problems and a few others as well. I call this the MacGyver method, after the old television show which showed MacGyver using whatever was at hand to cleverly solve his problem. The show celebrated a triumph of brain over brawn.

The MacGyver method is based on bivariate estimation of correlations. It assumes that the selected DCC model is correctly specified between every pair of assets i and j. Hence the correlation process is simply given by

$$\left. \begin{aligned} \rho_{i,j,t} &= \frac{q_{i,j,t}}{\sqrt{q_{i,i,t}q_{j,j,t}}}, \\ q_{i,j,t} &= \bar{R}_{i,j}(1 - \alpha - \beta) + \alpha s_{i,t-1}s_{j,t-1} + \beta q_{i,j,t-1}, \end{aligned} \right\} \tag{6.1}$$

and the log likelihood function for this pair of assets is easily extracted from (4.39). It is given by

$$L_{2,i,j} = -\frac{1}{2} \sum_t \left(\log[1 - \rho_{i,j,t}^2] + \frac{s_{i,t}^2 + s_{j,t}^2 - 2\rho_{i,j,t} s_{i,t} s_{j,t}}{[1 - \rho_{i,j,t}^2]} \right). \qquad (6.2)$$

Because the high-dimension model is correctly specified, so is the bivariate model. The estimates are maximum-likelihood estimates (MLEs) of the third step of the estimations. The volatility parameters and unconditional correlations are estimated as before. The only parameters are (α, β) and the estimates will be consistent using only data on one pair of assets. However, it is obvious that information is being ignored that could yield more efficient estimates. Thus improved estimation should combine parameter estimates from these bivariate models. The combined parameters are then used with equation (6.1) to calculate the correlations.

An analytical solution to the optimal combination of bivariate parameter estimates seems extremely difficult to find. The data are dependent from one pair to another and the dependence is a function of the parameters. Knowing the dependence of the data does not lead easily to measures of the dependence of the parameter estimates. Perhaps an analytical solution to this problem might arise at some point—in the meantime, I will develop the estimator based on Monte Carlo performance.

A variety of simulation environments is postulated. In each case all bivariate pairs are estimated and then simple aggregation procedures such as means or medians are applied. Several issues immediately arise. What should be done about cases where the estimation does not converge or where it converges to a value outside the region of stationarity? When averaging parameters that have constrained ranges, it is easy to introduce bias.

Six estimators will be considered. The mean, median, and trimmed mean are constructed from either an unrestricted maximization of (6.2) or from a restricted maximization. These will be called aggregators.

The trimmed means are computed by deleting the largest 5% and the smallest 5% of the estimates and then taking the mean of the remaining ones. The unrestricted MLE simply maximizes the log likelihood (6.2) without restriction. If it does not converge in a finite number of iterations, then the final value of the estimate is taken. Obviously, this estimation can, and occasionally does, lead to some very bizarre parameter estimates. The restricted MLE reparameterizes the log likelihood using a logistic functional form so that both parameters must lie in the interval $(0, 1)$. Their sum was not restricted in this case. The model is expressed

Table **6.1.**   Experiments for MacGyver simulation.

	Number	α	β	Rhobar
Exp1	3.00	0.05	0.90	0.50
Exp2	5.00	0.05	0.90	0.50
Exp3	10.00	0.05	0.90	0.50
Exp4	20.00	0.05	0.90	0.50
Exp5	30.00	0.05	0.90	0.50
Exp6	50.00	0.05	0.90	0.50
Exp7	10.00	0.05	0.94	0.50
Exp8	10.00	0.02	0.97	0.50
Exp9	10.00	0.05	0.90	0.20
Exp10	10.00	0.05	0.90	0.80

as

$$q_{i,j,t} = \bar{R}_{i,j} + \frac{e^{\theta}}{1+e^{\theta}}(s_{i,t-1}s_{j,t-1} - \bar{R}_{i,j}) + \frac{e^{\phi}}{1+e^{\phi}}(q_{i,j,t-1} - \bar{R}_{i,j}). \quad (6.3)$$

The optimizer chooses (θ, ϕ) but the estimated values of (α, β) are passed back to be averaged across bivariate pairs.

Ten experiments are run with various parameter values and dimensions. All have a time series sample size of one thousand observations and one hundred replications of each experiment. The dimensions range from $n = 3$ to $n = 50$. The ten experiments are defined in table 6.1. The true correlation matrix has all unconditional correlations equal to a number called Rhobar in (6.1). In each case the parameters are estimated by bivariate MLE or restricted bivariate MLE and then the summary measures are computed according to each aggregator. The ultimate result is a table of root mean squared (RMS) errors and a table of biases across the simulations for each of the two parameters, α and β.

The tables of RMS errors and biases are presented as tables 6.2 and 6.3. For each experiment, the estimator with the smallest RMS error is shown in bold. The net result is that the smallest errors are achieved by the median estimator. For β, the best estimator for each experiment is either the median or the median of the restricted estimator. On average, the median of the unrestricted estimator is the smallest. For α, the medians are best in most experiments and the median of the unrestricted bivariate parameter estimates has the smallest RMS error. This estimator effectively ignores all the nonconvergent and nonstationary solutions and gives parameter estimates that are very close to the true value.

The biases of these estimators are also of interest. In all experiments the bias in β is negative and the bias in α is positive. This is not surprising in a context where β is truncated from above (at 1) while α is truncated

Table 6.2. RMS errors from the MacGyver method.

	α mean unrestricted	α mean restricted	α trimmed mean unrestricted	α trimmed mean restricted	α median unrestricted	α median restricted
Exp1	**0.01092**	0.01233	0.01150	0.01386	0.01150	0.01386
Exp2	0.00720	0.00807	0.00710	0.00854	**0.00682**	0.00843
Exp3	0.02550	0.00543	0.00528	0.00504	**0.00491**	0.00510
Exp4	0.01509	0.00419	0.00438	0.00383	0.00410	**0.00357**
Exp5	5.63324	0.00403	0.00389	0.00361	0.00358	**0.00321**
Exp6	0.01088	0.00352	0.00301	0.00308	**0.00274**	0.00275
Exp7	0.00578	0.00380	0.00430	**0.00371**	0.00424	0.00379
Exp8	471.66375	0.00552	0.00377	0.00391	**0.00283**	0.00295
Exp9	10.58470	0.00469	0.00424	0.00432	**0.00402**	0.00441
Exp10	0.00554	0.00498	0.00542	**0.00483**	0.00539	0.00516
Average	48.79626	0.00566	0.00529	0.00547	**0.00501**	0.00532

	β mean unrestricted	β mean restricted	β trimmed mean unrestricted	β trimmed mean restricted	β median unrestricted	β median restricted
Exp1	0.03743	0.05134	0.03417	0.03584	**0.03417**	0.03584
Exp2	0.03693	0.03630	0.02716	0.02910	**0.02186**	0.02438
Exp3	1.20351	0.02736	0.02080	0.02040	0.01629	**0.01515**
Exp4	0.21512	0.02543	0.01796	0.01729	0.01289	**0.01179**
Exp5	0.07398	0.02388	0.01542	0.01592	**0.01041**	0.01047
Exp6	0.14473	0.02400	0.01490	0.01578	**0.00969**	0.01022
Exp7	0.01311	0.00819	0.00705	0.00747	**0.00578**	0.00703
Exp8	88.71563	0.05939	0.06618	0.03728	**0.01015**	0.01178
Exp9	15.72034	0.03100	0.02297	0.02220	0.01510	**0.01449**
Exp10	0.02063	0.01894	0.01585	0.01453	0.01361	**0.01177**
Average	10.61814	0.03058	0.02425	0.02158	**0.01499**	0.01529

Table 6.3. Bias from the MacGyver method.

	α mean unrestricted	α mean restricted	α trimmed mean unrestricted	α trimmed mean restricted	α median unrestricted	α median restricted
Exp1	0.00071	0.00110	0.00024	0.00094	0.00024	0.00094
Exp2	0.00124	0.00285	0.00077	0.00210	0.00079	0.00153
Exp3	-0.00036	0.00261	0.00184	0.00195	0.00082	0.00106
Exp4	0.00352	0.00254	0.00222	0.00188	0.00142	0.00088
Exp5	-0.55754	0.00248	0.00183	0.00180	0.00099	0.00095
Exp6	0.00096	0.00224	0.00152	0.00152	0.00071	0.00067
Exp7	0.00272	0.00158	0.00214	0.00134	0.00134	0.00092
Exp8	-48.11060	0.00428	0.00158	0.00298	0.00142	0.00174
Exp9	-1.05017	0.00251	0.00208	0.00192	0.00072	0.00113
Exp10	0.00247	0.00233	0.00215	0.00196	0.00183	0.00170
Average	-4.97070	0.00245	0.00164	0.00184	**0.00103**	0.00115

	β mean unrestricted	β mean restricted	β trimmed mean unrestricted	β trimmed mean restricted	β median unrestricted	β median restricted
Exp1	-0.01710	-0.02087	-0.01165	-0.01022	-0.01165	-0.01022
Exp2	-0.01945	-0.02216	-0.01450	-0.01674	-0.00942	-0.01079
Exp3	0.09886	-0.02275	-0.01615	-0.01605	-0.01036	-0.00933
Exp4	-0.01104	-0.02333	-0.01524	-0.01518	-0.00951	-0.00900
Exp5	-0.02231	-0.02257	-0.01333	-0.01429	-0.00741	-0.00811
Exp6	-0.00020	-0.02264	-0.01344	-0.01414	-0.00765	-0.00789
Exp7	0.00175	-0.00661	-0.00062	-0.00577	-0.00252	-0.00479
Exp8	11.18001	-0.05195	-0.01487	-0.02937	-0.00859	-0.00998
Exp9	1.54597	-0.02653	-0.01931	-0.01814	-0.01033	-0.00994
Exp10	-0.01524	-0.01525	-0.01106	-0.01087	-0.00803	-0.00666
Average	1.27412	-0.02347	-0.01302	-0.01508	**-0.00855**	-0.00867

below (at 0). Notice, however, that this bias is in the opposite direction from the bias observed by Engle and Sheppard (2005b), who found α to be too small and β to be too big for large systems. Also notice that the bias is very small. On average over the experiments, the bias in α is 0.001 and the bias in β is -0.008. Since these biases result from bivariate estimation, there is no large-system bias, as there is for MLE estimation of DCC. In fact, the RMS errors are smallest for the largest systems and the biases are generally reduced for large systems.

In addition to the computational simplification and bias reduction, there are several other advantages to this MacGyver method of estimating a DCC model. When there are 50 assets, there are 1,225 bivariate pairs. When there are 100 assets, there are 4,950 asset pairs. Hence the number of bivariate estimations increases as well. However, since only the median of all these estimations is needed, there is little loss of efficiency if some are not run. This opens up the possibility of estimating a subset of the bivariate pairs. While it is not clear how to select a good subset, it is clear that there is little advantage to computing all of them. When new assets are added to the collection, it may not be necessary to reestimate at all if the investigator is confident that the specification is adequate.

A second advantage is that the data sets for each bivariate pair need not be of the same length. Thus, an asset with only a short history can be added to the system without requiring the shortening of all other series. This is particularly important when examining large asset classes and correlations across countries as there are many assets that are newly issued, merged, or otherwise associated with short time histories.

A potential third advantage that will not be explored in this chapter is that there may be evidence in these bivariate parameter estimates that the selected DCC model is not correctly specified. The bivariate models would presumably show less dispersion if the model is correctly specified than if it is incorrect.

Recently, alternative approaches to approximating the log likelihood have been introduced. These approaches allow quasi-maximum-likelihood estimation to be performed without having to invert large matrices. In Engle et al. (2008b), a composite likelihood function is proposed that essentially sums the bivariate likelihoods so that only one estimation is required. The theoretical and empirical performance of this method appears to be very promising.

7

Generalized DCC Models

7.1 Theoretical Specification

The great advantage of DCC models is the parsimony of parameterization. For the simple mean-reverting DCC model with correlation targeting, there are only two unknown parameters in the correlation process, no matter how many variables are being modeled. The integrated DCC only has one and the asymmetric DCC has three. This parsimony is in direct contrast to the general versions of multivariate GARCH, where the number of parameters tends to rise with the square or cube of the number of assets. It has long been believed that it would not be possible to estimate large covariance matrices with multivariate GARCH models.

The DCC solution rests on some dramatic simplifications of the structure. First, it is assumed that univariate GARCH models are adequate for each of the series considered. This assumption is more restrictive than many multivariate GARCH specifications in that only the past of each series is used to model the variances. It is more general than most multivariate GARCH models in that the parameters, and indeed the structure, of each of the GARCH models can be different. One series can be modeled with an EGARCH model, while another might have a volatility series that is derived from option implied volatilities. The second assumption is that the correlation process is very simple and has the same dynamics for each pair of assets.

The question of whether these restrictions give adequate estimates of the covariance matrices is essentially an empirical one. What do the typical variables look like and are these restrictions supported in the data? Can we do diagnostic tests to reveal when data do not support these parameter restrictions?

In order to examine these restrictions, it is helpful to have a class of less restricted models. Cappiello et al. (2007) introduced the generalized DCC model. It has symmetric and asymmetric versions, both of which we will discuss. They also allow for a structural break in the model. The model, the tests, and the results will be discussed next.

There are, however, many other ways to generalize the correlation structure of the DCC. Pelletier (2006) and Billio and Caporin (2005) have a regime-switching model of DCC. Billio et al. (2006) introduce a block structure DCC model. Fernandes et al. (2005) build a model based on the range statistic. Hafner and Franses (2003) propose a variant on the generalized DCC model. Patton (2006a) points out the econometric issues involved when estimating correlations between many series with different numbers of observations. Silvennoinen and Terasvirta (2005) propose smooth transition changes in correlations.

The standard symmetric mean-reverting DCC model defined in (4.10) is written as the following two equations:

$$Q_t = \Omega + \alpha \varepsilon_{t-1} \varepsilon'_{t-1} + \beta Q_{t-1}, \tag{7.1}$$

$$R_t = \text{diag}(Q_t)^{-1/2} Q_t \, \text{diag}(Q_t)^{-1/2}. \tag{7.2}$$

The matrix R will be a correlation matrix as long as Q is positive definite. This is guaranteed in this case by the assumptions that $(\alpha, \beta, 1 - \alpha - \beta) > 0$ and that Ω is positive definite. More general models must also have this property.

The generalized DCC (GDCC) model replaces equation (7.1) with

$$Q_t = \Omega + A' \varepsilon_{t-1} \varepsilon'_{t-1} A + B' Q_{t-1} B. \tag{7.3}$$

This model is clearly still positive definite but has $2n^2$ parameters plus the intercept. This increase in flexibility increases the estimation requirements dramatically, making large systems impractical. Perhaps, however, not all the parameters in the matrices A and B are needed to adequately fit the data.

In Cappiello et al. (2007) a diagonal specification is proposed for A and for B. This introduces a separate value of α and β for each asset and restricts the correlation dynamics to depend upon the product of these two. Let a be an $n \times 1$ vector and set $A = \text{diag}(a)$ to be a matrix with the elements of a on the diagonal and zeros elsewhere. Similarly, set $B = \text{diag}(b)$. There are now $2n$ parameters plus the intercept. A typical element of Q can be expressed as

$$Q_{i,j,t} = \Omega_{i,j} + a_i a_j \varepsilon_{i,t-1} \varepsilon_{j,t-1} + b_i b_j Q_{i,j,t-1}. \tag{7.4}$$

Some assets may have relatively constant correlations with small values for a. Others may fluctuate dramatically with each new data point, in which case a would be large.

A similar structure can be applied to the asymmetric models. In its simplest form the equation for Q is

$$Q_t = \Omega + \alpha \varepsilon_{t-1} \varepsilon'_{t-1} + \gamma \eta_{t-1} \eta_{t-1} + \beta Q_{t-1}, \quad \eta_t = \min(\varepsilon_t, 0). \tag{7.5}$$

This is therefore called the asymmetric DCC (ADCC) model. It has only one additional parameter. The generalized version of this model is

$$Q_t = \Omega + A'\varepsilon_{t-1}\varepsilon'_{t-1}A + G'\eta_{t-1}\eta_{t-1}G + B'Q_{t-1}B. \tag{7.6}$$

If A and B are diagonal and G is also taken to be diagonal, with elements given by the vector g, then the typical element is

$$Q_{i,j,t} = \Omega_{i,j} + a_i a_j \varepsilon_{i,t-1} \varepsilon_{j,t-1} + g_i g_j \eta_{i,t-1} \eta_{j,t-1} + b_i b_j Q_{i,j,t-1}. \tag{7.7}$$

This is the asymmetric generalized DCC (AGDCC) model, which is the model estimated by Cappiello et al. (2007).

One of the computational advantages of the DCC models is their ability to use correlation targeting to replace the intercept parameters with a consistent estimate. This becomes more complicated in these cases and may not be worth the effort in some models. Consider the average of Q from (7.6):

$$\bar{Q} \equiv \frac{1}{T} \sum_{t=1}^{T} Q_t$$

$$= \Omega + \frac{1}{T} A' \sum_{t=1}^{T} (\varepsilon_{t-1}\varepsilon'_{t-1})A + \frac{1}{T} G' \sum_{t=1}^{T} (\eta_{t-1}\eta'_{t-1})G + B'\bar{Q}B. \tag{7.8}$$

If we suppose that $\bar{Q}$ is the same as the average correlation, $\bar{R}$, then

$$\Omega = \bar{R} - A'\bar{R}A - G'\bar{N}G - B'\bar{R}B, \tag{7.9}$$

where

$$\bar{R} \equiv \frac{1}{T} \sum_{t=1}^{T} (\varepsilon_{t-1}\varepsilon'_{t-1}) \quad \text{and} \quad \bar{N} \equiv \frac{1}{T} \sum_{t=1}^{T} (\eta_{t-1}\eta'_{t-1}).$$

This equation can be solved for any set of parameter values and for any data set. To keep Q in (7.6) positive definite, it is sufficient to keep Ω positive definite. However, the constraints needed to keep Ω positive definite are hard to define and can lead to numerical difficulties when maximizing the likelihood.

For a proposal on how to approximate this matrix, see Hafner and Franses (2003) and Hafner et al. (2006).

Cappiello et al. (2007) wanted to test whether the covariance matrix changed after the introduction of the euro. This is a specific date on which it was hypothesized that the parameters might have changed. There are several ways in which the parameters could have changed: the variance parameters might have changed, dynamic parameters might have changed, and the intercept, Ω, might have changed. Each of these changes can be expressed by an interaction with a dummy variable that is set to 1 after the introduction of the euro.

To allow a break in the intercept, the model can be expressed as

$$Q_t = (\bar{R}_1 - A'\bar{R}_1 A - B'\bar{R}_1 B - G'\bar{N}_1 G)(1 - d_t)$$
$$+ (\bar{R}_2 - A'\bar{R}_2 A - B'\bar{R}_2 B - G'\bar{N}_2 G)d_t$$
$$+ A'\varepsilon_{t-1}\varepsilon'_{t-1}A + G'\eta_{t-1}\eta'_{t-1}G + B'Q_{t-1}B. \qquad (7.10)$$

In this case, separate estimates of both the full correlation matrix and the negative part of the correlation matrix are needed before and after the break.

If, in addition, the parameters are different before and after, the model can be expressed as

$$Q_t = (\bar{R}_1 - A'\bar{R}_1 A - B'\bar{R}_1 B - G'\bar{N}_1 G)(1 - d_t)$$
$$+ (\bar{R}_2 - A'_2\bar{R}_2 A_2 - B'_2\bar{R}_2 B_2 - G'_2\bar{N}_2 G_2)d_t$$
$$+ [A'\varepsilon_{t-1}\varepsilon'_{t-1}A + G'\eta_{t-1}\eta'_{t-1}G + B'Q_{t-1}B](1 - d_t)$$
$$+ [A'_2\varepsilon_{t-1}\varepsilon'_{t-1}A_2 + G'_2\eta_{t-1}\eta'_{t-1}G_2 + B'_2 Q_{t-1}B_2]d_t. \quad (7.11)$$

Needless to say, the number of parameters increases rapidly when structural breaks are introduced. However, tests of subsets of these conditions can be implemented by specializing the relations in equation (7.11).

7.2 Estimating Correlations for Global Stock and Bond Returns

The data employed in Cappiello et al. (2007) consist of FTSE All-World indices for twenty-one countries and DataStream-constructed five-year-average-maturity bond indices for thirteen. This gives a total of thirty-four asset returns. The sample covers the period from January 8, 1987, until February 7, 2002—a total of 785 observations. All returns were continuously compounded using Thursday-to-Thursday closing prices to avoid any end-of-week effects. Each of the assets was DE-GARCHED by choosing the best of nine different GARCH model specifications using the Bayesian Information Criterion, or BIC. As can be seen in table 7.1, most of the bond returns use the simple symmetric GARCH model, while most of the equity returns achieve their best fits using one of the asymmetric models. The most useful is the ZGARCH, and this is followed by the EGARCH and TARCH.

Four different correlation models are estimated: the simple scalar DCC, the generalized DCC, the asymmetric scalar DCC, and the asymmetric generalized DCC. These are denoted DCC, GDCC, ADCC and AGDCC. Each of these four models is also estimated with a break in the mean

Table 7.1. Selected GARCH models (with asymmetric ones in bold).

Method	Assets selecting method
GARCH	3EQ, 8BOND
AVGARCH	0
NGARCH	1BOND
EGARCH	6EQ, 1BOND
ZGARCH	8EQ, 1BOND
TARCH	3EQ, 1BOND
APARCH	0
AGARCH	1EQ, 1BOND
NAGARCH	0

Table 7.2. Log likelihood for various models. (The following abbreviations are made in the table: "WMB" denotes "with mean break" and "WMDB" denotes "with mean and dynamics breaks.")

Model	LL value	Number of parameters in the correlation evolution	Approximate BIC
DCC	$-25,722.1$	$561 + 2$	-1.057
DCC WMB	$-24,816.2$	$1,122 + 2$	-0.743
DCC WMDB	$-24,789.2$	$1,122 + 4$	-0.742
GDCC	$-25,564.5$	$561 + 68$	-0.853
GDCC WMB	$-24,572.5$	$1,122 + 68$	-0.723
GDCC WMDB	$-24,483.3$	$1,122 + 136$	-0.708
ADCC	$-25,704.7$	$561 + 3$	-0.869
ADCC WMB	$-24,809.2$	$1,122 + 3$	-0.742
ADCC WMDB	$-24,781.6$	$1,122 + 6$	-0.741
AGDCC	$-25,485.1$	$561 + 102$	-0.844
AGDCC WMB	$-24,487.5$	$1,122 + 102$	-0.714
AGDCC WMDB	$-24,398.3$	$1,122 + 204$	-0.694

of the correlation process and with a break in both the mean and the dynamic parameters of the process. The results are tabulated in table 7.2.

As can be seen from the table there are many parameters being estimated. The parameter count includes the parameters of the correlation intercept matrix even though they are estimated by correlation targeting. Thus the simplest model has a parameter count of $561 + 2$, or 563, parameters. Adding an asymmetric term to get the ADCC improves the

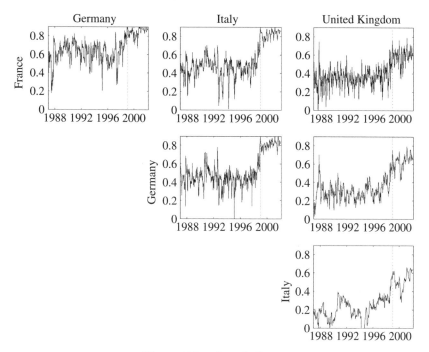

Figure 7.1. Correlations.

BIC. The generalized DCC (GDCC) also improves the BIC relative to DCC. Adding a break in the mean doubles the intercept parameters, but this is still selected as an improvement in BIC. In fact, of all the models, the preferred one is the most highly parameterized: the AGDCC with breaks in the mean and the dynamics.

The parameter estimates are presented for the GDCC and AGDCC model without breaks in table 7.3. It is apparent here that the parameters are similar, but not identical, across countries. For most equities, the asymmetric term is bigger than the symmetric term, while the opposite is true for most bond returns.

Figure 7.1 shows a small part of the correlation matrix. It shows the equity correlations between France and Germany, France and Italy, and France and the United Kingdom across the top. All three are clearly rising but perhaps the rise is a little greater between France and Germany and between France and Italy. The growth of Europe and its increasing integration are presumably part of the explanation for the increases. As the companies in each country become more international, it is natural that they would respond to shocks in a similar fashion. A second cause for the increase may be increasing global financial volatility. It is widely recognized that volatilities and correlations move together, so

Table 7.3. Parameter estimates. (An asterisk after a figure in the table means that that coefficient is not significantly different from zero at the 5% level.)

	Symmetric model		Asymmetric model		
	a_i^2	b_i^2	a_i^2	g_i^2	b_i^2
Australian stocks	0.0002*	0.9641	0.0062	0.0078	0.7899
Austrian stocks	0.0084	0.9481	0.0032	0.0042	0.9606
Belgian stocks	0.0139	0.9490	0.0104	0.0081	0.9501
Canadaian stocks	0.0066	0.9186	0.0051	0.0024	0.9523
Danish stocks	0.0077	0.9468	0.0034	0.0052	0.9646
French stocks	0.0094	0.9438	0.0086	0.0027	0.9454
German stocks	0.0122	0.9448	0.0071	0.0066	0.9568
Hong Kong stocks	0.0022	0.9655	0.0004*	0.0022	0.9563
Irish stocks	0.0045	0.9647	0.0002*	0.0067	0.9677
Italian stocks	0.0135	0.9488	0.0071	0.0117	0.9569
Japanese stocks	0.0026	0.9497	0.0020	0.0026	0.9526
Mexican stocks	0.0012	0.9635	0.0009*	0.0189	0.9375
Dutch stocks	0.0099	0.9562	0.0061	0.0091	0.9587
New Zealand stocks	0.0000*	0.9574*	0.0010*	0.0009*	0.9215
Norwegian stocks	0.0076	0.9235	0.0017	0.0057	0.9290
Singaporean stocks	0.0013	0.9492	0.0006*	0.0021	0.9760
Spanish stocks	0.0090	0.9463	0.0055	0.0073	0.9538
Swedish stocks	0.0075	0.9676	0.0049	0.0055	0.9649
Swiss stocks	0.0118	0.9542	0.0145	0.0092	0.9427
U.K. stocks	0.0079	0.9484	0.0066	0.0064	0.9549
U.S. stocks	0.0090	0.9261	0.0020	0.0040	0.9512
Austrian bonds	0.0131	0.9759	0.0096	0.0087	0.9762
Belgian bonds	0.0168	0.9712	0.0112	0.0089	0.9745
Canadian bonds	0.0077	0.9418	0.0053	0.0056	0.8593
Danish bonds	0.0186	0.9678	0.0111	0.0090	0.9731
French bonds	0.0146	0.9721	0.0106	0.0079	0.9734
German bonds	0.0167	0.9712	0.0131	0.0090	0.9715
Irish bonds	0.0161	0.9700	0.0138	0.0065	0.9675
Japanese bonds	0.0087	0.9627	0.0047	0.0063	0.9642
Dutch bonds	0.0166	0.9714	0.0132	0.0076	0.9716
Swedish bonds	0.0119	0.9618	0.0081	0.0117	0.9615
Swiss bonds	0.0138	0.9754	0.0117	0.0067	0.9740
U.K. bonds	0.0091	0.9689	0.0058	0.0041	0.9716
U.S. bonds	0.0096	0.9277	0.0058	0.0027	0.9361
Scalar model	0.01010	0.94258	0.00817	0.00653	0.94816

this effect should lead to increasing correlations. A third explanation is the asymmetry in correlations. As markets turned down after 2000, the negative returns naturally increased the correlations. Hence some of the increase following 2000 may be attributable to the asymmetric correlation models. Finally, the date at which the exchange rate is fixed is shown as a dashed line on the graphs. There is clearly an increase in correlations at that time, even between the United Kingdom and countries that adopted the euro. Thus the fixing of the euro exchange rates may be part of the story but it is not the full story.

Cappiello et al. (2007) present a model of correlations between thirty-four financial assets. Many interesting dynamic patterns are examined in the paper but only a few have been discussed here. This is only the beginning of what is possible with the DCC model. The generalizations that have been introduced can help to make the models more realistic while retaining feasible computational complexity.

8
FACTOR DCC

8.1 Formulation of Factor Versions of DCC

To model large numbers of asset returns, the profession has always turned to factor models. This is a natural development as the investigator seeks to discover the small number of important factors that influence a large number of returns. The Capital Asset Pricing Model (CAPM) of Sharpe (1964) is a one-factor model of all asset returns. It is the primary model behind most financial decision making because it delivers a powerful theory on the pricing and hedging of financial assets. It was then extended to the Arbitrage Pricing Theory (APT) by Ross (1976). These path-breaking frameworks have been the workhorses for most asset pricing models ever since.

These structures have had less influence in the modeling of covariance matrices. The difficulty is that many factors may be needed and the βs that connect factors with returns are difficult to estimate precisely and are unlikely to be constant over long periods of time. Thus there is some appeal to using time series methods to get the benefits of a factor structure without needing such careful formulation of the number and measurement of the factors.

In general there are two classes of factor models: models with observable factors and models with unobservable factors. In the first case, the investigator specifies the factor or factors and then attempts to estimate the covariance matrix of returns with these factors. In the second case, he must first learn about the factors from the data and then build a model that incorporates this information. The goal in each case is to find forecasting models to predict the covariance matrix one or more periods in the future. This prediction will generally require predicting the distribution of the factor and then using this to predict the covariance matrix of the rest of the returns. The role of the factors in this analysis is generally to give a parsimonious representation of the correlation structure that is at the same time guaranteed to be positive definite. The FACTOR DCC model, which will be introduced below, has both observable and unobservable factors. The unobservable factors are assumed to be only temporarily important for explaining dynamic correlations.

Consider first the very simple static one-factor model that is the centerpiece of the CAPM. Measuring returns in excess of the risk-free rate and letting r_{m} be the market return, the model is most simply expressed as

$$r_{i,t} = \alpha_i + \beta_i r_{\mathrm{m},t} + \varepsilon_{i,t}. \tag{8.1}$$

From theoretical arguments, we expect the αs to be 0 in an efficient market. And we expect the idiosyncratic returns to be uncorrelated across assets:

$$\left.\begin{aligned}
V(r_{i,t}) &= \beta_i^2 V(r_{\mathrm{m},t}) + V(\varepsilon_{i,t}), \\
\mathrm{Cov}(r_{i,t}, r_{j,t}) &= \beta_i \beta_j V(r_{\mathrm{m},t}).
\end{aligned}\right\} \tag{8.2}$$

Thus the correlation between two assets can be expressed as

$$\rho_{i,j} = \frac{\beta_i \beta_j V(r_{\mathrm{m},t})}{\sqrt{(\beta_i^2 V(r_{\mathrm{m},t}) + V(\varepsilon_{i,t}))(\beta_j^2 V(r_{\mathrm{m},t}) + V(\varepsilon_{j,t}))}}. \tag{8.3}$$

These expressions ensure that the correlation matrix will be positive definite. They do not, however, provide any measures of time-varying variances, covariances, or correlations.

The simplest approach to formulating a dynamic version of this one-factor model is to follow Engle et al. (1990b) and Ng et al. (1992). In this case the factor has time-varying volatility and can be modeled with some form of ARCH model. Consequently, the expressions in (8.2) and (8.3) can be rewritten in terms of conditional variances. The conditional correlation then become

$$\rho_{i,j,t} = \frac{\beta_i \beta_j V_{t-1}(r_{\mathrm{m},t})}{\sqrt{(\beta_i^2 V_{t-1}(r_{\mathrm{m},t}) + V_{t-1}(\varepsilon_{i,t}))(\beta_j^2 V_{t-1}(r_{\mathrm{m},t}) + V_{t-1}(\varepsilon_{j,t}))}}. \tag{8.4}$$

The model used by Engle et al. assumed that the idiosyncratic volatilities were not changing over time. They called this the FACTOR ARCH model and I will use that name here. Letting β and r be $n \times 1$ vectors, the statistical specification is

$$V_{t-1}\begin{pmatrix} r_t \\ r_{\mathrm{m},t} \end{pmatrix} = \begin{pmatrix} \beta\beta' h_{\mathrm{m},t} + D^2 & \beta h_{\mathrm{m},t} \\ \beta' h_{\mathrm{m},t} & h_{\mathrm{m},t} \end{pmatrix}, \quad D \sim \text{diagonal}. \tag{8.5}$$

The conditional correlation between each pair of assets would be time varying only because the market volatility is changing. From an examination of (8.5) it is clear that the conditional correlation in this model is a monotonic function of market volatility ranging from 0 to 1 as market volatility ranges from 0 to ∞.

The expression for the conditional covariance matrix of returns and factors given in (8.5) implicitly assumes that this matrix is nonsingular. However, in many cases the market return is an exact linear combination of individual asset returns. The weights in this linear combination are time varying, so it is well-known that there is a small logical inconsistency in models such as this. Of course, if only a subset of the returns in the index is being modeled, there is no longer an overlooked constraint and the model is logically coherent. We will assume this is the case although in reality, the importance of this singularity is minuscule for the correlation estimation problem.

In the FACTOR ARCH model, there will always be portfolios of assets that have no ARCH. Using the common-features approach of Engle and Kozicki (1993), Engle and Susmel (1993) looked for such portfolios and found that they unlikely in an international context. Almost all portfolios have time-varying volatility, even if they have a zero β on the market. Hence, there must either be more factors or time-varying idiosyncratic volatility.

Assuming normal errors, the statistical model is

$$r_t | r_{m,t}, \mathbb{F}_{t-1} \sim N(\beta r_{m,t}, D^2), \qquad r_{m,t} | \mathbb{F}_{t-1} \sim N(0, h_{m,t}). \qquad (8.6)$$

The MLE of this model is simply a regression of each asset return on the market return with ordinary least squares (OLS), and an MLE of the market volatility by GARCH. There is no benefit to system estimation.

A natural extension of this model is to allow the idiosyncrasies, as well as the market return, to follow a GARCH process. Thus there are two GARCH processes for an asset. For convenience we will call this model a FACTOR DOUBLE ARCH. The model is expressed as

$$V_{t-1} \begin{pmatrix} r_t \\ r_{m,t} \end{pmatrix} = \begin{pmatrix} \beta\beta' h_{m,t} + D_t^2 & \beta h_{m,t} \\ \beta' h_{m,t} & h_{m,t} \end{pmatrix}, \quad D_t \sim \text{diagonal GARCH}, \quad (8.7)$$

where D_t is a diagonal matrix with GARCH standard deviations on the diagonal. Assuming conditionally normal returns this can be rewritten as

$$r_t | r_{m,t}, \mathbb{F}_{t-1} \sim N(\beta r_{m,t}, D_t^2), \qquad r_{m,t} | \mathbb{F}_{t-1} \sim N(0, h_{m,t}). \qquad (8.8)$$

This model is still easy to estimate by MLE. The return on an asset is regressed on the market return with disturbances that follow a GARCH. The GARCH for the market is then only estimated once. To see that this two-step estimator is an MLE, express the likelihood for this problem as the density of asset returns conditional on the market return times the marginal density of market returns. Ignoring irrelevant constants, the log likelihood is given by

$$L(r, r_{\mathrm{m}}) = - \sum_{t=1}^{T} \log |D_t| - \frac{1}{2} \sum_{t=1}^{T} (r_t - \beta r_{\mathrm{m},t}) D_t^{-2} (r_t - \beta r_{\mathrm{m},t})$$

$$- \frac{1}{2} \sum_{t=1}^{T} \left[\log(h_t) + \frac{r_{\mathrm{m},t}^2}{h_t} \right]. \quad (8.9)$$

This model satisfies the weak exogeneity conditions of Engle et al. (1983), which allow for separate estimation of the conditional and marginal models. As long as the parameters are distinct or *variation free*, so that no information from the marginal model affects inference in the conditional model, then market returns can be considered to be weakly exogenous and the MLE of the system is the same as the MLE done in two steps.

There are many reasons to believe that the one-factor DOUBLE ARCH model just described will still be too simple to accurately forecast correlations. The correlations between stock returns in the same industry are typically higher than for stocks across industries, and these correlations will rise if the industry volatility rises. These correlations arise from additional factors with impacts on correlations that vary over time. Even more interesting are factors that have zero variance some of the time and a large variance at other times. Energy prices might fall into this category. It would be impossible to identify this factor until it became active, but by then then it may be too late. Finally, the model assumes that the factor loadings or βs are constant over time, yet whenever a firm changes its line of business, its sensitivity to various factors will naturally change.

Ideally, the model should allow correlations among idiosyncrasies and between idiosyncrasies and market shocks, and these correlations should be time varying. In this way the statistical model will recognize the changing correlation structure when a new factor emerges or factor loadings change.

The FACTOR DCC model is designed to do just this. It proceeds exactly as described above for the FACTOR DOUBLE ARCH and then estimates a DCC model on the residuals. More precisely, the FACTOR DCC model has the specification

$$r_t = \beta r_{\mathrm{m},t} + D_t \varepsilon_t, \quad r_{\mathrm{m},t} = \sqrt{h_t} \varepsilon_{\mathrm{m},t}, \quad \begin{pmatrix} \varepsilon_t \\ \varepsilon_{\mathrm{m},t} \end{pmatrix} \sim N(0, R_t). \quad (8.10)$$

The specification of the correlation matrix can be the same as for any of the DCC models discussed earlier in this book. Partitioning the

$(n + 1) \times (n + 1)$ correlation matrix into its conformal parts as

$$R_t = \begin{pmatrix} R_{1,1,t} & R_{1,m,t} \\ R_{m,1,t} & 1 \end{pmatrix}, \tag{8.11}$$

the covariance matrix of returns is given by

$$\left. \begin{aligned} V_{t-1}(r_t) &= \beta\beta' h_{m,t} + D_t R_{1,1,t} D_t + \sqrt{h_{m,t}} (\beta R_{m,1,t} D_t + D_t R_{1,m,t} \beta'), \\ V_{t-1}\begin{pmatrix} r_t \\ r_{m,t} \end{pmatrix} &= \begin{pmatrix} V_{t-1}(r_t) & \beta h_{m,t} + \sqrt{h_{m,t}} D_t R_{1,m,t} \\ \beta' h_{m,t} + \sqrt{h_{m,t}} R_{m,1,t} D_t & h_{m,t} \end{pmatrix}. \end{aligned} \right\} \tag{8.12}$$

When new factors arise, the correlations between some stocks will increase. These changes are captured by the second term in the upper-left block. When βs change, these effects will be captured by the third and fourth terms. This model specifies one observable factor but allows many unspecified unobservable factors that are assumed to obey the DCC dynamics and thus are mean reverting. If returns are jointly normal, then equation (8.12) gives the covariance matrix. No longer will returns conditional on the market have a linear regression coefficient. Instead

$$r_t | r_{m,t}, \mathbb{F}_{t-1} \sim N([\beta + h_{m,t}^{-1/2} D_t] r_{m,t}, W_t). \tag{8.13}$$

The terms in square brackets are the time-varying βs and the covariance matrix W is the same as the upper-left block of (8.12) minus $\beta_t \beta_t' h_{m,t}$.

The model in equation (8.10) is only a small generalization of the basic DCC model. The data in this case are not just standardized returns but standardized idiosyncratic returns. If either the FACTOR ARCH or the FACTOR DOUBLE ARCH are correctly specified, then the DCC should predict zero correlations both conditionally and unconditionally. Estimation is naturally done in two steps again, where the first step estimates both the static factor loading and the idiosyncratic GARCH—the second step estimates the DCC parameters.

The conditional correlations are again defined as the conditional covariance divided by the product of the conditional standard deviations using the expression for the conditional covariance matrix of returns in (8.12). In each case there are now four terms and the last three depend on the DCC estimated correlations.

This one-factor version of FACTOR DCC is easily generalized to a multifactor setting. If there are k observable factors and these are given by $(f_{1,t}, f_{2,t}, \ldots, f_{k,t})' = F_t$, then the distribution of returns conditional on the factors in the DOUBLE ARCH model is expressed by

$$r_t | F_t, \mathbb{F}_{t-1} \sim N(BF_t, D_t^2), \qquad F_t | \mathbb{F}_{t-1} \sim N(0, G_t). \tag{8.14}$$

The matrix of factor loadings B is an $n \times k$ matrix. In this case the factors are assumed to have a multivariate GARCH specification with a $k \times k$ time-varying covariance matrix G. Of course this again assumes that idiosyncrasies are uncorrelated and that the factor loadings are constant over time.

The k-factor DCC model then allows the residuals of (8.14) to have a DCC specification. This is now an $n + k$ dimension vector of residuals, which can be expressed as

$$\varepsilon_{r,t} = D_t^{-1}(r_t - BF_t), \qquad \varepsilon_{f,t} = G_t^{-1/2} F_t, \tag{8.15}$$

$$V_{t-1}\begin{pmatrix} \varepsilon_{r,t} \\ \varepsilon_{f,t} \end{pmatrix} = \begin{pmatrix} R_{r,r,t} & R_{r,f,t} \\ R_{f,r,t} & G_t \end{pmatrix}. \tag{8.16}$$

The conditional covariance matrix of returns can now be expressed as

$$\begin{aligned} V_{t-1}(r_t) &= E_{t-1}(BF_t + D_t\varepsilon_t)(BF_t + D_t\varepsilon_t)' \\ &= BG_t B' + D_t R_{r,r,t} D_t + BR_{f,r,t} D_t + D_t R_{r,f,t} B'. \end{aligned} \tag{8.17}$$

This expression can be used to define conditional correlations and to define the joint Gaussian likelihood just as above.

The empirical implementation of the k-factor DCC model is slightly more complex than the one-factor version because it is now necessary to have a multivariate GARCH model of the factors. This could itself be a DCC. There is, however, an additional reason why this method is complicated. When multiple factors are considered, there are many βs involved. If a relatively unimportant factor is considered, it may have insignificant βs for many assets. However, some of these βs could be large, with large standard errors. In such a case, the covariances will incorporate this information in ways that introduce noise. Consequently, it may be that some shrinkage of the βs would be useful. Factors will help in correlation estimation if they are significant in some of the return equations. They do not need to be in all the equations, so perhaps they should be set to 0. Under what criterion should a factor be modeled? Is it enough for it to be significant in a small number of assets? The answers to these questions are unclear. This complexity informally suggests the importance of the flexible specification in the FACTOR DCC.

8.2 Estimation of Factor Models

To examine the properties of these correlation estimators a set of eighteen daily U.S. large-cap equity returns will be examined. The data comprise 2,771 observations from 1994 to 2004. The symbols for these stocks are aa, axp, ba, cat, dd, dis, ge, gm, ibm, ip, jnj, jpm, ko, mcd, mmm, mo, mrk, and msft, which are all components of the Dow Jones Industrial Average. The S&P 500 is taken as the market return.

8.2.1 MacGyver Estimates

The MacGyver method is applied to this data set to estimate all the correlations with DCC. Although it is not necessary to use the same GARCH model for each series, I do just that in this investigation. To account for the asymmetry in volatility, the GJR or threshold GARCH model is used. It is specified by

$$r_{i,t} = \sqrt{h_{i,t}}\varepsilon_{i,t}, \quad h_{i,t} = \omega_i + \theta_i r_{i,t-1}^2 + \gamma_i r_{i,t-1}^2 I_{r_{i,t}\ _1<0} + \phi h_{i,t-1}. \quad (8.18)$$

The standardized returns from these models are saved and used as inputs for the DCC estimation by MacGyver. For eighteen returns there are $18 \times \frac{17}{2} = 153$ bivariate models. For most of these, the results are quite standard. For a few, though, they are completely unsatisfactory. For example, all of the αs are estimated to be between 0 and 0.05 except for one that is just over 2. Similarly, most of the βs are less than 1 but for the same bivariate estimate, β is over 4,000. A few of the βs are quite small or negative. Nevertheless, the medians are very close to the values we generally find. The median for α is 0.0157 and the median for β is 0.9755, so the sum is just over 0.99, which leads to a good degree of persistence in correlations.

The DCC estimation produces 153 time series of correlations based on these two parameters and the unconditional correlations. It is difficult to examine so many time series at once. Some clear patterns can easily be seen by looking at the average correlations. These will establish the stylized facts of correlations in the U.S. equity market. In figure 8.1, the mean correlation is plotted from the 100-day historical method and the DCC method with TARCH volatilities.

The historical correlations and the DCC correlations trace out very much the same pattern. The range of the historical correlations is a little greater but this may be a result of the choice of smoothing. A 200-day correlation would move substantially less. The historical correlations also have wider peaks, which makes the correlation estimate somewhat slower to respond to news. A plot of the cross-sectional standard deviation of the 153 bivariate correlations reveals that the historical correlations are more varied across pairs than the DCC.

It is clear that these correlations have changed substantially over the ten-year period. The highest correlations are during the recession in 2002 and the first part of 2003. Correlations are low during the Internet

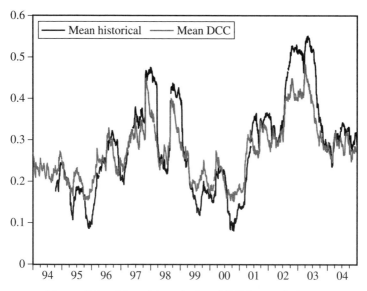

Figure 8.1. Mean historical and DCC correlations.

bubble and the subsequent bursting of the bubble. They rise in 2001 and abruptly increase further after 9/11. In 2003 correlations fall as the economy and stock market recover. There are two episodes of spiking correlations in the late 1990s; these are associated with the LTCM/Russian default crisis and the Asian currency crisis. In fact, the proximate cause of the second spike is the "Anniversary crash" on October 27, 1997, when the market fell 7% and then recovered 5% the next day. These events are plotted with the correlations in figure 8.2.[1] It certainly appears to be the case that economic crises lead to rising correlations.

8.2.2 FACTOR ARCH, FACTOR DOUBLE ARCH

The sharp movements in correlations that are associated with movements in the S&P 500 itself suggest the usefulness of a factor model. The FACTOR ARCH and the FACTOR DOUBLE ARCH are now calculated. They follow the specification (8.5) and (8.7). The βs are estimated by OLS for the FACTOR ARCH and by generalized least squares (GLS) with GARCH errors for the FACTOR DOUBLE ARCH and they are consequently slightly different. From figure 8.3 it is clear that these differences are small for all eighteen stocks.

---

[1] The LTCM dummy is defined for August 1998 through September 25, 1998; the Asian currency crisis dummy is defined for May 14, 1997, through July 31, 1997; the Anniversary crash dummy is defined for October 27 and October 28, 1997.

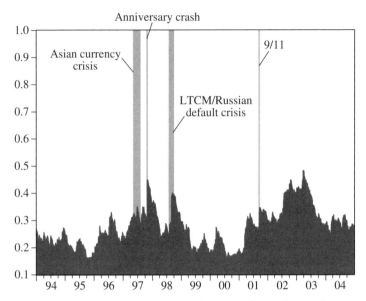

Figure 8.2.   Mean correlations of DCC and significant dates.

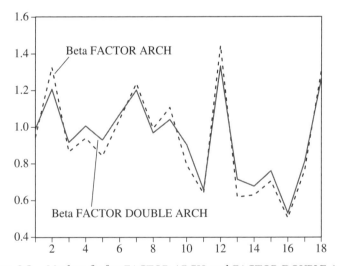

Figure 8.3.   Market βs for FACTOR ARCH and FACTOR DOUBLE ARCH.

The correlations from each of these models can be calculated using (8.4). The average across all pairs is again a useful measure. This is shown in figure 8.4.

The average correlation from the FACTOR DOUBLE ARCH model is very similar in level to that in the average DCC model. It differs primarily in that the FACTOR DOUBLE ARCH correlations are more volatile. When

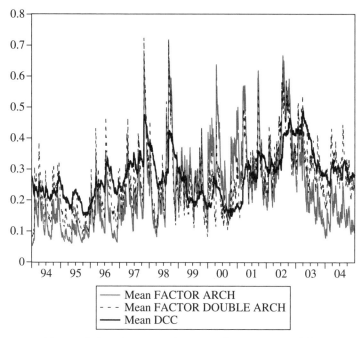

Figure 8.4. Mean correlations of FACTOR models.

the correlations spike upward because of some market event, they rise
to 0.7 in several cases, and when the correlations fall, they fall further. It
is not clear whether the higher volatility is a good or bad aspect of this
estimator as we do not know what the true conditional correlations are
at any point in time.

The patterns of the FACTOR ARCH are, however, different in several
important ways. Over the last two years of the sample, the FACTOR
ARCH correlations fall much lower than they do for any of the other
correlation estimators. This is also the case in the mid 1990s. The opposite occurs in 1999 and 2000, when the FACTOR ARCH correlations are
higher than for DCC and FACTOR DOUBLE ARCH. These differences are
easy to understand. The monotonic relationship between average correlation and market volatility in the FACTOR ARCH model implies that the
correlations should be at their lowest in the mid 1990s and after 2003,
since market volatility is lowest at these times. However, the idiosyncratic volatilities also change generally in the same direction. The most
accurate correlation estimates—either from the DCC model or from the
FACTOR DOUBLE ARCH model—reduce the movements in correlations.
During the Internet bubble, the opposite effect is observed. The market
volatility is high but so are idiosyncratic volatilities so the correlations

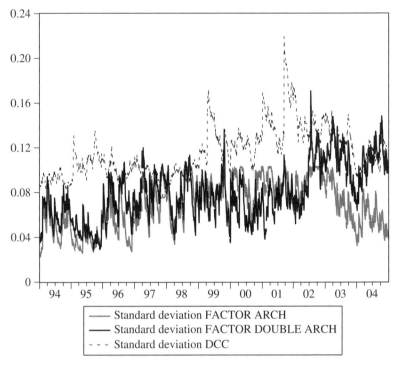

Figure 8.5. Cross-sectional standard deviations of factor correlations.

are low. The FACTOR ARCH model cannot model this. The observation of Campbell et al. (2001) that idiosyncratic volatilities are rising should not be interpreted as a trend but rather as a process that ultimately reverses in about 2002.

The cross-sectional standard deviations of these three estimators are interesting. They are shown in figure 8.5.

From the figure it is clear that the DCC model has correlations that differ more across pairs than the two factor models. Perhaps this is not surprising as the component due to the factor is the same for all pairs in the FACTOR models, whereas each pair has its own time series in the DCC. Thus the FACTOR DOUBLE ARCH model is more volatile over time but the DCC is more volatile cross-sectionally. It remains to be seen whether this feature of the models is a good or a bad thing.

8.2.3 FACTOR DCC

As discussed above, FACTOR DCC simply estimates a DCC model from the residuals of the FACTOR DOUBLE ARCH model following the specification in (8.10). The MacGyver method is used to estimate the parameters of this DCC. The median $\alpha = 0.009$, while the median $\beta = 0.925$.

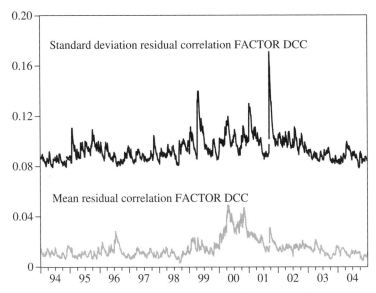

Figure 8.6. Mean and standard deviation of residual correlations from FACTOR DCC.

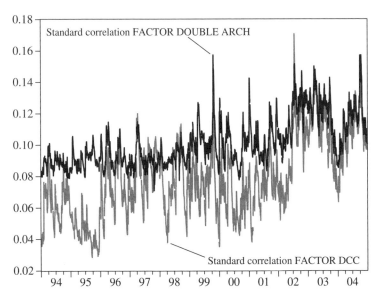

Figure 8.7. Cross-sectional standard deviation of FACTOR correlations.

The sum of these two numbers is much farther from unity than the DCC estimates on the simple returns; the correlation process is therefore less persistent. Because α is smaller, it is also less volatile.

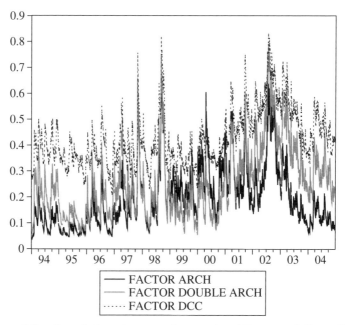

Figure 8.8.   Correlations between International Paper and Caterpillar.

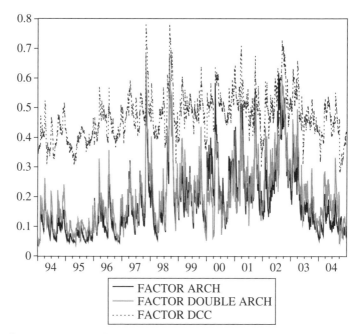

Figure 8.9.   Correlations between Merck and Johnson & Johnson.

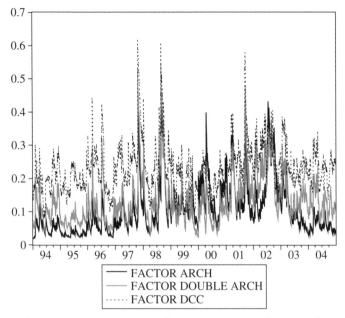

Figure 8.10. Correlations between Altria and Coca-Cola.

The residuals from the FACTOR DOUBLE ARCH model should be uncorrelated both conditionally and unconditionally if the one-factor model is correct; the DCC on these residuals might not find anything. The average residual correlation and its cross-sectional standard deviation are shown in figure 8.6.

The average correlation of the residuals is quite small. It averages 0.01 over cross-sectional pairs. It does rise in the middle of the sample but only to 0.04. The cross-sectional standard deviation is, however, of the same order of magnitude as the cross-sectional standard deviation of the DCC, although it does not rise quite so much. Thus the average correlation is small but it has substantial cross-sectional variability.

When these residual correlations are incorporated into the calculation of the conditional correlations, the result is a substantial change for some pairs and very little for many others. In fact, the average correlation looks almost identical to the FACTOR DOUBLE ARCH. However, the cross-sectional dispersion is now greater. The cross-sectional standard deviation of FACTOR DCC is shown in figure 8.7.

The reasons for these differences are easily seen in a few examples shown in figures 8.8–8.10. Stocks in the same industry have idiosyncratic shocks that are correlated. The FACTOR DCC method incorporates these idiosyncratic correlations into the correlation estimates. If these residual correlations are constant, the correction is static, but if it is dynamic, then a time-varying correction is automatically generated by the FACTOR DCC method. Figures 8.8–8.10 show the correlations estimated between International Paper and Caterpillar, between Merck and Johnson & Johnson, and between the Coca-Cola Company and Phillip Morris.

9

Anticipating Correlations

9.1 Forecasting

The goal of this book is to develop methods to anticipate correlations. Thus far, the book has developed descriptions of correlations and why they change as well as models that can track the changes. The task of anticipating correlations seems far more formidable. However, every model that has been presented predicts correlations one period in the future. This short-horizon prediction will be sufficient for many financial applications. The measurement of Value-at-Risk and high-frequency hedging demands may rely on this forecast horizon. The most important feature of short-horizon forecasts is the updating. As new information becomes available, the models will systematically update the volatilities and correlations to produce an estimate of the next outcome. A model which does not update volatilities and correlations will make much bigger mistakes when the markets are changing.

However, multistep predictions of correlations are more complex and may be very important for further analyses. A natural goal is an unbiased forecast of correlations. This cannot be achieved for a very simple reason. The correlation is always between $+1$ and -1. An unbiased estimator must have some probability of being above the true value as well as a probability of being below it. If the true correlation were very close to 1, then an estimator could only miss on the downside and would be biased.

If unbiased forecasts are not what is required, then what is? Statistical decision theory provides the answer that it depends on the losses that occur if bad parameter estimates are used. The best forecast of the correlation would be the one that minimizes the expected losses. This problem has not yet been solved as far as I know.

For many financial applications, the covariance is the quantity that needs to be forecast in addition to the variances. For example, to predict the variance of a portfolio, the variances of the components and the covariances between components would both be needed. Thus if unbiased forecasts of these elements could be calculated, then an unbiased estimate of portfolio variance would be achieved. Risk measures are based on this statistic. Hedge ratios depend on ratios of covariances and variances. Such hedge ratios are the solution to a problem such as

$$\min_{\beta} E_t(y_{1,t+k} - \beta y_{2,t+k})^2. \tag{9.1}$$

It is well-known that the answer is

$$\hat{\beta} = \frac{E_t(y_{1,t+k} y_{2,t+k})}{E_t(y_{2,t+k}^2)}. \tag{9.2}$$

Thus it is the forecast of both covariances and variances that is required to get an optimal hedge ratio. If the hedge is to be held over several periods, the numerator and the denominator of (9.2) will each be summed over the hedge period.

The formulation of the multistep forecasting problem that seems most relevant is

$$\hat{H}_{t+k/t} = E_t(y_{t+k} - E_t y_{t+k})(y_{t+k} - E_t y_{t+k})'. \tag{9.3}$$

These multistep forecasts can then be translated into correlation forecasts. Such forecasts will not be unbiased estimates of correlations but will be the ratio of unbiased estimates of covariances and the associated estimates of variances. For assets 1 and 2, this gives

$$\rho_{t+k/t} = \frac{\hat{H}_{12,t+k/t}}{\sqrt{\hat{H}_{11,t+k/t} \hat{H}_{22,t+k/t}}}. \tag{9.4}$$

In the FACTOR ARCH model, it is possible to see exactly how this works. In this case unbiased estimates of the multistep forecasts of the covariance can be computed but forecasts of the correlations for the same two assets would be based on (8.3). This expectation is a function of only a single statistic: the market, or factor, variance. If this is high, then correlations will be high:

$$\rho_{t+k,t} = \frac{\beta_1 \beta_2 h_{t+k/t}}{\sqrt{(\beta_1^2 h_{t+k/t} + V(\varepsilon_1))(\beta_2^2 h_{t+k/t} + V(\varepsilon_2))}} \neq E_t(\rho_{t+k}). \tag{9.5}$$

Forecasting correlations for this model primarily involves forecasting market variance k steps ahead. Macroeconomic factors could enter the analysis here, much as they do in Engle et al.'s (2008b) Spline GARCH model discussed in the next section. If the economy is likely to have more inflation or a recession, market volatility would increase; if the economy is forecast to have stable growth, then financial market volatility would decrease.

In the DCC model, the criterion (9.4) for the forward correlation between assets 1 and 2 becomes

$$\rho_{t+k,t} = \frac{E_t(\rho_{t+k}\sqrt{h_{1,t+k}h_{2,t+k}})}{\sqrt{E_t(h_{1,t+k})E_t(h_{2,t+k})}}, \tag{9.6}$$

where

$$\left.\begin{array}{l} \rho_{t+k} = \dfrac{Q_{1,2,t+k}}{\sqrt{Q_{1,1,t+k}Q_{2,2,t+k}}}, \\[3mm] Q_{t+k} = \bar{R}(1 - \alpha - \beta) + \alpha\varepsilon_{t+k-1}\varepsilon'_{t+k-1} + \beta Q_{t+k-1}. \end{array}\right\} \tag{9.7}$$

Note that for zero-mean processes, as before,

$$\begin{aligned} E_t(y_{1,t+k}y_{2,t+k}) &= E_t(\sqrt{h_{1,t+k}h_{2,t+k}}\varepsilon_{1,t+k}\varepsilon_{2,t+k}) \\ &= E_t(\rho_{t+k}\sqrt{h_{1,t+k}h_{2,t+k}}). \end{aligned} \tag{9.8}$$

If $k = 1$, then the expectation is unnecessary. For $k > 1$ it is clear that there is no analytical expectation for the numerator of (9.6). A Taylor series expansion of the numerator gives

$$\begin{aligned} \rho_{t+k}&\sqrt{h_{1,t+k}h_{2,t+k}} \\ &\cong \hat{\rho}_{t+k}\sqrt{h_{1,t+k/t}h_{2,t+k/t}} + \sqrt{h_{1,t+k/t}h_{2,t+k/t}}(\rho_{t+k} - \hat{\rho}_{t+k}) \\ &\quad + \hat{\rho}_{t+k}\sqrt{\frac{h_{2,t+k/t}}{4h_{1,t+k/t}}}(h_{1,t+k} - h_{1,t+k/t}) \\ &\quad + \hat{\rho}_{t+k}\sqrt{\frac{h_{1,t+k/t}}{4h_{2,t+k/t}}}(h_{2,t+k} - h_{2,t+k/t}) + w, \end{aligned} \tag{9.9}$$

where w is a remainder. Taking expectations of this expression gives

$$\begin{aligned} E_t(\rho_{t+k}&\sqrt{h_{1,t+k}h_{2,t+k}}) \\ &= \hat{\rho}_{t+k}\sqrt{h_{1,t+k/t}h_{2,t+k/t}} + \sqrt{h_{1,t+k/t}h_{2,t+k/t}}(E_t(\rho_{t+k}) - \hat{\rho}_{t+k}) + E(w) \\ &= E_t(\rho_{t+k})\sqrt{h_{1,t+k/t}h_{2,t+k/t}} + E(w), \end{aligned} \tag{9.10}$$

since the forecasts of conditional variance are unbiased. Substituting into (9.6) gives

$$\rho_{t+k/t} = E_t(\rho_{t+k}) + \frac{E(w)}{\sqrt{h_{1,t+k/t}h_{2,t+k/t}}}. \tag{9.11}$$

Thus the best forecast of the conditional correlation will be the expected correlation as long as the first-order Taylor approximation is sufficiently accurate.

In order to calculate $E_t(\rho_{t+k})$, a further approximation is necessary. This is because it is not possible to analytically forecast Q or R because

$$E_t Q_{t+k} = \bar{R}(1 - \alpha - \beta) + \alpha E_t R_{t+k-1} + \beta E_t Q_{t+k-1}. \tag{9.12}$$

Engle and Sheppard (2005b) consider several solutions. In their analysis these are all similar in performance and consist of assuming that

$$E_t R_{t+j} = E_t Q_{t+j}. \tag{9.13}$$

If the diagonal elements of Q are near unity, then the nonlinearity is unimportant. Forecasts of Q will have diagonal elements that approach 1 so this approximation may be especially good for large j. Substituting (9.13) into (9.12) gives

$$E_t R_{t+k} = \bar{R} + (\alpha + \beta)(E_t R_{t+k-1} - \bar{R}). \tag{9.14}$$

Correlations are forecast to gradually mean revert from R_{t+1} to the long-run average with a time constant that depends on $(\alpha + \beta)$, just as for conventional GARCH(1, 1) processes.

To evaluate the remainder term of (9.10), the next terms of the Taylor expansion can be examined. Evaluating the second-order terms gives an improved expression:

$$
\begin{aligned}
&E_t(\sqrt{h_{1,t+k} h_{2,t+k}}) \\
&\cong \sqrt{h_{1,t+k/t} h_{2,t+k/t}} \\
&\quad - \frac{\sqrt{h_{1,t+k/t} h_{2,t+k/t}}}{8} \left\{ \frac{V_t(h_{1,t+k})}{h_{1,t+j/t}^2} + \frac{V_t(h_{2,t+k})}{h_{2,t+j/t}^2} - 2\frac{\mathrm{Cov}_t(h_{1,t+k}, h_{2,t+k})}{h_{1,t+k/t} h_{2,t+k/t}} \right\}.
\end{aligned}
\tag{9.15}
$$

This expression confirms that for multistep forecasts there will be an upward bias in the plug-in estimate of the covariance from DCC models in (9.10) even if the correlation is accurately predicted. Intuitively, the bias due to taking the square root and the bias from taking the product of the expectations offset completely only in the case where the conditional variances are perfectly correlated.

To evaluate these approximations, Engle and Sheppard simulate the DCC process. For a particular set of parameter values and starting conditions they simulate 1,000 paths of 1,022 days each. They estimate the DCC process on the first 1,000 observations and then use three forecast approximations for the remaining 22 days. Deviations from the forecast and the true simulated correlations incorporate both parameter uncertainty and the forecasting approximation. In general, the errors are small, do not depend upon the linearization, and decrease with horizon and with the size of the underlying unconditional correlation. The solution in (9.14) thus seems like a reasonable approach for computing multistep correlation forecasts. Obviously, the heart of the task remains the one-step forecast.

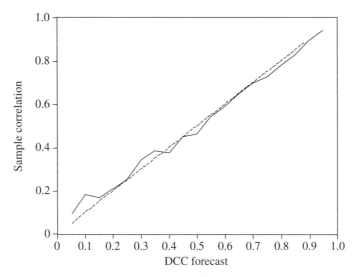

Figure 9.1. The 100-day DCC forecast and simulated
sample correlations (- - -, the 45 degree line).

To further evaluate these approximations without parameter uncertainty, the DCC model is simulated 100,000 times, fixing the parameters at values that are used later in this chapter. That is, the DCC parameter α is 0.0157 and β is 0.9755. The GARCH parameters are estimated parameters for General Electric and have values $(0.037\,07, 0.961\,48)$ showing a high degree of persistence. Initially, the correlations are started at the steady-state value so that all horizons are predicted to have this value. The return correlation at the 100-day horizon is then compared with this forecast. In figure 9.1, the forecast and actual correlations at 100 days are shown for various underlying correlation structures. It is clear that the predicted correlations are quite close to the expected values.

In a second simulation, the DCC process is started with correlations of 0 but a long-run correlation of 0.5. Over many periods the correlations are forecast to gradually rise to the level of 0.5. As a comparison, the correlation between returns over the 100,000 simulations can be compared. These are called RFORECAST1 and RACTUAL1. A similar comparison starts the correlations at 0.9 and allows them to gradually decline to 0.5. These are labeled RFORECAST2 and RACTUAL2. Both these comparisons are shown in figure 9.2. The pattern of the forecasts and the actual correlations are very similar, although for long horizons there is an upward bias of perhaps 0.05 in the correlation forecast. This is a result of the approximations made in (9.11), (9.13) and seems to be consistent with the expression (9.15). These appear to be quite small in the context of forecasts several years in the future.

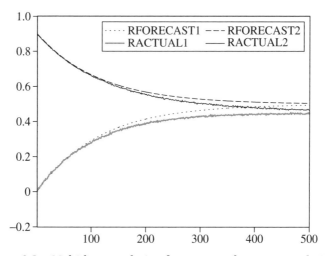

Figure 9.2. Multiday correlation forecasts and return correlations.

   In the DCC model, correlations change over time in response to shocks in returns. Thus the investor cannot expect future correlations to be the same as today's correlations even if that is the best forecast of the model. There is a distribution of possible future correlations that can be used for assessing risk. This confidence interval for correlations can be determined by simulation. In figure 9.3 histograms are plotted for correlations after 10 days, 50 days, and 500 days. These are denoted RHO_10, RHO_50, and RHO_500. As can be seen, the range of correlations that is expected increases with horizon until it reaches a stationary distribution. In this case, even though the correlation is forecast to be 0.5 for the indefinite future in this example, the 95% confidence interval after 10 days for the forward one-day correlation is $(0.426, 0.566)$. Over 100 days this interval becomes $(0.322, 0.635)$. Thus correlations can change and our measures of risk should incorporate this possibility.

9.2   Long-Run Forecasting

The need to forecast correlations and volatilities over many periods arises naturally in financial applications. Many portfolio decisions are designed for the long run and derivatives are often written and hedged with a long-run perspective. The models described in this book are all mean-reverting ones and, consequently, have relatively uninteresting long-run forecasts. The idea that the volatility of all assets will mean revert to their mean level and that correlations will revert to their mean value implies that, at least for long-run decision making, a static

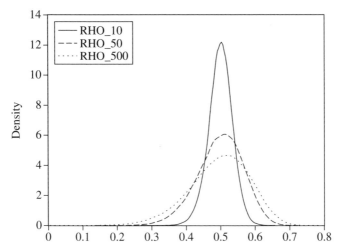

Figure 9.3. Kernel densities of correlations after 10, 50, and 500 days.

approach should be sufficient. The dynamic model could suggest how long it would take to get to this level but the level is determined by the historical sample being used.

In the FACTOR models the key variable to forecast is the volatility of the market. Since this is typically specified as a GARCH, the long-run forecast of correlations in this class of models is simply based on the long-run forecast of market volatility. Again, the mean reversion in this factor is a starting point but a better approach is needed.

Recently, a new set of volatility and correlation models have been introduced that break this simple relation. The first models are the Spline GARCH model of Engle et al. (2008b). In this model, a simple extension to a univariate GARCH model has interesting forecasting implications. The model is formulated so that the volatility is the product of a slowly varying component and a unit GARCH. Thus, for an asset with return y,

$$y_t = \sqrt{\tau_t g_t}\varepsilon_t, \quad g_t = 1 - \alpha - \beta + \alpha\frac{y_{t-1}^2}{\tau_{t-1}} + \beta g_{t-1}, \\ \log(\tau_t) = \text{quadratic spline.} \qquad (9.16)$$

Thus the level of volatility can evolve gradually. The model requires us to specify the type of spline and the number of knots and then the system can be estimated jointly. For medium-run forecasting, only the spline will matter as the GARCH will mean revert to unity. However, for long-run forecasting it is not clear whether to extrapolate the spline, keep it at its final value, or seek some other explanation for its future movement.

Instead, Engle and Rangel (2008) propose modeling the spline with macroeconomic variables that determine why equity market volatility is sometimes low and sometimes high. They fit the Spline GARCH model to about fifty countries for up to fifteen years and then model the annual values of the spline as a function of macroeconomic and financial variables as well as global common effects. The results are quite intuitive. The economic effects that give rise to higher long-run volatility are

- high inflation,

- slow output growth and recession,

- high volatility of short-term interest rates,

- high volatility of output growth,

- high volatility of inflation.

As each of these factors contributes to higher volatility of a market return, they are also long-run predictors of correlations.

Thinking about the U.S. economic environment in the fall of 2007 and at the beginning of 2008, these factors all seem to be relevant. It is not clear whether the macroeconomy is headed for recession or inflation, or both. It is clear that sort-term interest rates have been adjusted very rapidly. These factors are presumably part of the reason why financial volatility was high in the second half of 2007.

In a second paper, Rangel and Engle (2008) use the Spline GARCH to model the factor in a one-factor model. Just as the stationary FACTOR ARCH model can be dramatically improved by allowing the idiosyncrasies to have time-varying volatilities, they can have nonstationary volatilities modeled by Spline GARCH. And in addition, the residuals can again be modeled with a DCC. This new class of factor models is called the FACTOR Spline GARCH model. They show striking evidence for the importance of these spline models within the factor structure, particularly in terms of long-run forecasts.

An alternative approach to the Spline GARCH is a MIDAS GARCH, as proposed by Engle et al. (2008a). In this model the long-run component is either constructed from realized volatilities or constructed directly from macroeconomic observations taking account of the different data-sampling frequencies.

A model designed to estimate long- and short-run correlations directly has been proposed by Colacito et al. (2008). This model is like DCC except that there is an intercept in the multivariate model that is a long weighted average of daily variances and covariances. Thus the long-run variances and covariances are nonstationary and therefore not mean reverting. The short-run DCC mean reverts to these long-run values.

9.3 Hedging Performance In-Sample

This book has reviewed and introduced many models for forecasting correlations, but do they work? To establish which of these models does the best job of forecasting correlations, we need to be clearer on what we want the forecasts to do for us. A natural criterion is based on portfolio optimization or hedging. If we have a better forecast of correlations, then we can form more efficient portfolios. This is an example of the methodology introduced by Engle and Colacito (2006) and is discussed in chapter 2. The optimal portfolio of two stocks with equal expected return is the minimum-variance combination.

The minimum-variance combination of assets (i, j) is given by

$$\left. \begin{array}{l} r_{\text{port},t} = w_t r_{i,t} + (1 - w_t) r_{j,t}, \\[2mm] w_t = \dfrac{h_{j,j,t} - h_{i,j,t}}{h_{i,i,t} + h_{j,j,t} - 2h_{i,j,t}}, \qquad V_{t-1}(r_t) = H_t. \end{array} \right\} \qquad (9.17)$$

Thus the optimal proportion of each asset to hold is changing over time based on the forecast of the covariance matrix. To achieve this optimal holding, the investor would forecast the next-day covariance matrix just before the close and then adjust his portfolio to have the weights given in (9.17). The criterion for success is that the portfolio has a smaller variance than if the weights had not been changed. Practically, unless this benefit covers transaction costs, we would not use it. However, for the purpose here, we simply want to know which method of forecasting the covariance matrix achieves the lowest variance.

A closely related problem is that of holding a position in one stock because it has an abnormal expected return and hedging the position with a second stock. Typically, this would mean shorting the second stock to obtain a hedge portfolio with the minimum variance. Although the problem is different, the same approach can be used to solve it. The optimal hedge is given by

$$r_{\text{port},t} = r_{i,t} - \beta_{i,j,t} r_{j,t}, \quad \beta_{i,j,t} = h_{i,j,t}/h_{j,j,t}. \qquad (9.18)$$

The criterion for success is again simply the smallest variance of the portfolio.

These two criteria are applied for each of the models we have discussed to the data set of eighteen large-cap stocks from 1994 to 2004. In addition to the DCC, FACTOR ARCH, DOUBLE FACTOR ARCH, and FACTOR DCC, results are computed for a constant-covariance matrix constructed from the whole sample period and for a 100-day rolling covariance matrix called "100-day historical." The average annualized

Table 9.1. The average volatility of optimized portfolios.

	Optimal constant weights	100-day historical	DCC	FACTOR ARCH	DOUBLE ARCH	FACTOR DCC
Minimum variance	25.065 11	25.290 70	24.8603	25.032 71	24.846 35	24.819 57
Hedge	30.205 85	30.321 68	29.9712	30.269 49	30.041 15	29.925 96

volatility for each pair is averaged over all pairs to obtain a single number for the performance of a particular correlation estimator. The results are given in table 9.1. Naturally, the minimum-variance hedge produces a lower volatility than a long–short hedge. The average volatility of these large-caps over the period is 41%, so all hedges substantially reduce this.

The results show that for both criteria, the FACTOR DCC model produces the best hedge portfolio. For the hedging problem, the DCC is next, followed by the FACTOR DOUBLE ARCH, while for the minimum-variance criterion, the order is reversed. All except the 100-day historical volatility outperform the optimal constant set of weights and, in the hedging problem, the FACTOR ARCH. The differences are very small, however. It appears that the gains from a better model may be only a 1% reduction in volatility. This does not mean, however, that for other problems the gains will also be small.

To determine whether these differences are just random chance, we can look at how many of the pairs preferred one estimator to another. These winning percentages tell a much stronger story. Tables 9.2 and 9.3 show the fraction of times the row method beats the column method. The best method has the largest fractions in the labeled row. For example, in hedging, the FACTOR DCC is superior to a constant hedge for 84% of the pairs and superior to the DCC for 74%. It beats the historical hedge for 99% of the pairs, the FACTOR ARCH for 95%, and the DOUBLE ARCH for 72%. Although the differences are small, they are systematic.

9.4 Out-of-Sample Hedging

These results are very encouraging for the model. However, they demonstrate hedging performance on the same sample that was used to estimate the model. Thus it may not fare as well in a real-time risk environment. On the other hand, there are many thousands of observations on each series and the models have only a few parameters, so the ability of the models to overfit is limited. It is useful to examine the performance

Table 9.2. Fraction of minimum-variance portfolios where row beats column.

	CONST	HIST100	FACTOR DCC	DOUBLE ARCH	FACTOR ARCH	DCC
CONST	—	0.817	0.261	0.484	0.209	0.235
HIST100	0.183	—	0.007	0.144	0.013	0.013
DCC	0.739	0.993	—	0.765	0.444	0.307
FACTOR ARCH	0.516	0.856	0.235	—	0.163	0.209
DOUBLE ARCH	0.791	0.987	0.556	0.837	—	0.366
FACTOR DCC	0.765	0.987	0.693	0.791	0.634	—

Table 9.3. Fraction of hedges where row beats column.

	CONST	HIST100	FACTOR DCC	DOUBLE ARCH	FACTOR ARCH	DCC
CONST	—	0.693	0.16	0.654	0.261	0.163
HIST100	0.307	—	0.007	0.405	0.118	0.007
DCC	0.84	0.993	—	0.908	0.592	0.261
FACTOR ARCH	0.346	0.595	0.092	—	0.056	0.046
DOUBLE ARCH	0.739	0.882	0.408	0.944	—	0.275
FACTOR DCC	0.837	0.993	0.739	0.954	0.725	—

of these models on a new post-sample data set. This is particularly interesting as the low-volatility regime in the United States more or less came to an end in the summer of 2007. In the spring and early summer there were rumors that subprime mortgages would be a big problem, but the equity market did nothing spectacular until the end of July. On July 24 and 26 the market fell 2% each day, followed by a 3% drop on August 3, reaching a low on August 15. In less than a month it fell almost 10% and market volatilities soared. A model that can perform well in an out-of-sample period that is different from the fitting period is a particularly good model. This period is challenging.

Not only did aggregate volatility patterns change abruptly in the summer of 2007, but the comovements of stocks went in some extraordinary new directions. During this fall in the market, many hedge funds suffered big losses and anecdotal evidence attributes some of the price movements directly to hedge fund liquidations. Khandani and Lo (2007) give a detailed description of these events, including the liquidation hypothesis. Funds that held long–short positions found that others with the same positions were liquidating them so that the long positions were falling and the shorts rising. Stocks were moving in very peculiar directions. The biggest days for losses of these quantitative long–short strategies were

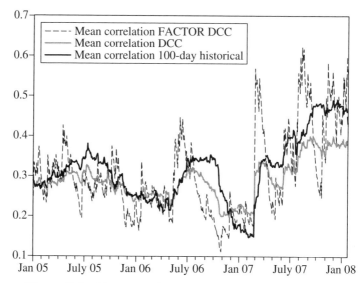

Figure 9.4. Mean correlations in the post-sample period.

August 7–9. One fund described these days as 25σ events. That is, the market moves were twenty-five times their standard deviation—an event that would happen only once in 10^{136} years for a normal distribution. As the fall season unfolded, normal markets did not return. Increasing discussion of recessions and hidden subprime losses further raised market volatilities and correlations.

In order to investigate the performance of the models through this period, new data were collected through January 23, 2008, from http://finance.yahoo.com. All the parameters of the models were fixed at their values estimated from 1994–2004 and the volatilities and correlations were updated to the end of the data set. The updates include volatilities and correlations of all eighteen large-caps plus the S&P 500.

Graphs of the mean correlations from several models show that the correlations continue to evolve and are rising dramatically at the end of the sample (see figure 9.4).

The DCC correlations are more slowly varying than the FACTOR DCC ones and are similar in level but not timing to the 100-day historical correlations. The interesting feature is the gradually falling mean correlation until the end of 2006 followed by the rise in correlations in 2007. The rise is punctuated by the big worldwide market response to the Chinese transaction tax on February 27, which temporarily spiked correlations, as measured by the FACTOR DCC, from about 15% to 50%. By July 2007, the mean correlation had risen to 35% according to all the models. Thereafter, it rose further still, reaching 50% in January 2008.

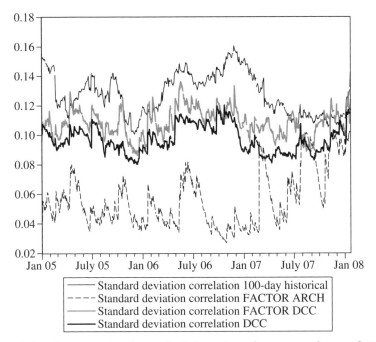

Figure 9.5. Cross-sectional standard deviation of post-sample correlations.

The high volatilities of the financial market in the second half of 2007 are also reflected in rising correlations.

Interestingly, the cross-sectional standard deviation of these correlations has come down by some measures, as shown in figure 9.8. In the middle of 2006, the cross-sectional dispersion of correlations was very low in the FACTOR ARCH model and very high in the 100-day historical correlations. As market volatility increased at the end of 2007, these both converged to the levels observed in DCC and FACTOR DCC.

From these estimates of volatilities and correlations, we can perform the same hedging and portfolio experiments as in the estimation sample. In table 9.4 the average volatility of minimum-variance pairs of portfolios and minimum-variance hedged long–short portfolios are constructed for the post-sample period.

In many respects this table tells the same story as before. The best estimators are still the DCC and FACTOR DCC, although the DCC is slightly better in the post-sample period for both criteria. The worst estimators are again the constant, 100-day historical, and FACTOR ARCH. For one criterion, the constant weight is the worst choice, while for the other, the FACTOR ARCH is the worst.

The comparison of winning percentages is similar but less dramatic, as can be seen in tables 9.5 and 9.6.

Table 9.4. Volatilities of optimized portfolios 2005–8.

	CONST	HIST100	FACTOR DCC	DOUBLE ARCH	FACTOR ARCH	DCC
Minimum variance	17.09745	16.17326	16.06358	17.00648	16.24782	16.13296
Hedge	20.0347	20.14098	19.91464	20.17249	19.96572	19.91764

Table 9.5. Fraction of pairs of minimum-variance portfolios where row beats column 2005–8.

	CONST	HIST100	FACTOR DCC	DOUBLE ARCH	FACTOR ARCH	DCC
CONST	—	0.255	0.163	0.412	0.235	0.203
HIST 100	0.745	—	0.32	0.712	0.536	0.431
DCC	0.837	0.68	—	0.83	0.732	0.654
FACTOR ARCH	0.588	0.288	0.17	—	0.229	0.222
DOUBLE ARCH	0.765	0.464	0.268	0.771	—	0.314
FACTOR DCC	0.797	0.569	0.346	0.778	0.686	—

Table 9.6. Fraction of pairs of long–short portfolios where row beats column 2005–8.

	CONST	HIST100	FACTOR DCC	DOUBLE ARCH	FACTOR ARCH	DCC
CONST	—	0.546	0.265	0.663	0.324	0.284
HIST 100	0.454	—	0.18	0.637	0.337	0.281
DCC	0.735	0.82	—	0.837	0.552	0.526
FACTOR ARCH	0.337	0.363	0.163	—	0.209	0.183
DOUBLE ARCH	0.676	0.663	0.448	0.791	—	0.408
FACTOR DCC	0.716	0.719	0.474	0.817	0.592	—

Using either criterion, the DCC forecast is the best for more than half the pairs, although it only marginally beats the FACTOR DCC. It beats the three weak estimators, CONST, HIST100, and FACTOR ARCH, in between 68% and 84% of the cases. This is a slightly stronger result than for FACTOR DCC. The DOUBLE ARCH model is in between, with good performance against the weak methods and weak performance against the two best methods.

It is important to recognize that this is a limited experiment and that the results may not be significantly different. It is not clear how to calculate statistical significance as the pairs are not at all independent. It does seem clear that the models do a decent job of estimating volatilities and correlations far out of sample and in dramatically different market environments.

9.5 Forecasting Risk in the Summer of 2007

Another interesting application of these models is to the assessment of risk. If the models are relatively stable over time, then they can be used on a daily basis to assess the risk of various portfolios even in turbulent market times. A plot that conveys this idea shows historical returns and plus and minus three times the predictive standard deviation. In this case, whenever the return exceeds this level, a 3σ event has occurred. For a normal distribution, a 3σ event should occur every year or two. We generally find that the normal tails are too thin for financial data, so we should expect somewhat more frequent occurrences of 3σ events. Risk managers must, on a daily basis, describe the risk environment of many different portfolios and typically use some calculation of standard deviation or Value-at-Risk as a measure. If the calculated standard deviation is too low, then the portfolio has more risk than would be chosen by a well-informed manager. It is interesting to speculate to what extent underestimation of risk contributed to the massive losses by financial institutions in the second half of 2007.

For the S&P 500, a confidence interval for returns can be constructed using parameters estimated through 2004. In this case it is simply an asymmetric GARCH model called TARCH (for threshold ARCH). The plot is shown in figure 9.6.

In this turbulent year there are only two events that would be classified as 3σ events. One is February 27, when China initiated a world market collapse, and the second is October 19, 2007. The confidence intervals are clearly increasing in August and during the period from October through to the end of the sample. The increase is sufficient that there are no major surprises.

Were the events of August therefore apparent to these models? The smallest confidence interval in July is on the 20th. The market declines on July 24, July 26, and August 2 lead to sharply increased bands so that the big market declines on August 3 and August 9 are not big surprises. Thus, from a one-step-ahead point of view, these events were not surprising. However, the succession of negative market returns moved the volatilities rapidly upward in a way that could not be predicted but that is consistent with models of this kind and that must be recognized in modern risk management. A multiperiod risk measure should account

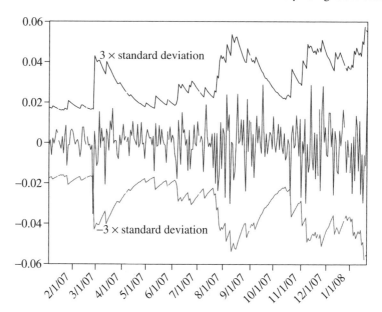

Figure 9.6. Returns and predicted standard deviations for the S&P 500.

for the unlikely but possible succession of returns in forming something like a Value-at-Risk. This is discussed by Engle (2004).

The same picture can be constructed for any portfolio. For example, a long–short portfolio that is constructed to reflect value or growth or momentum strategies might better reflect the risks faced by hedge funds through this period. A portfolio with weights on each of the stocks in this data set will have a variance given by

$$
\left.
\begin{aligned}
r_t^{\text{port}} &= \sum_{i=1}^{N} w_{i,t} r_{i,t}, \qquad h_{i,t} = V_{t-1}(r_{i,t}), \\
V_{t-1}(r_t^{\text{port}}) &= \sum_{i=1}^{N} w_{i,t}^2 h_{i,t} + 2 \sum_{i=1}^{N} \sum_{j>i}^{N} w_{i,t} w_{j,t} \sqrt{h_{i,t} h_{j,t}} \rho_{i,j,t}.
\end{aligned}
\right\}
\tag{9.19}
$$

Here are three examples of such portfolios. The first, called the value portfolio, is long an equal-weighted portfolio of stocks that underperformed the S&P 500 over the last year and short a portfolio that outperformed. The second follows the contrarian strategy of Lehmann (1990), Lo and MacKinlay (1990b), and Khandani and Lo (2007) by being long yesterday's losers in proportion to the size of the losses and short yesterday's winners. The third uses a CRSP measure of book to market in January 2007 to form a portfolio that is long stocks that have relatively high book to market and short those with low book to market. For each

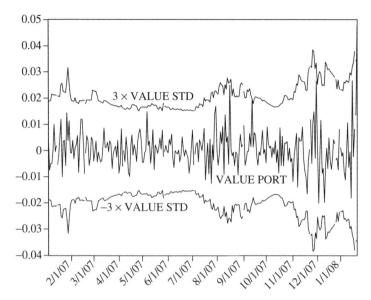

Figure 9.7. 3σ bands for value portfolio based
on previous-year underperformance.

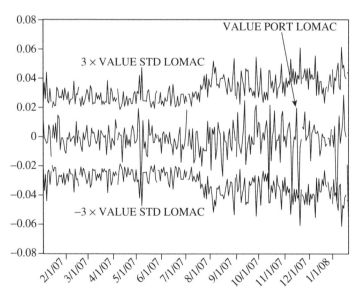

Figure 9.8. 3σ bands for Lo and MacKinlay
portfolio of yesterday's losers minus winners.

of these three portfolios of the eighteen large-cap stocks in this study, the portfolio return and variance are constructed as in (9.19). The results are summarized in figures 9.7–9.9.

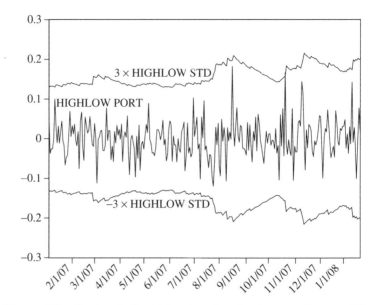

Figure 9.9. 3σ bands for high minus low book to market portfolio.

In the first figure, there is only one 3σ day and that is the last day in the sample (January 23, 2008), when this particular portfolio goes up by more than 4%. There is no evidence that this strategy would have been surprised by the August results. In fact, the confidence intervals gradually increase starting at the beginning of July so that the returns in August are easily within bounds on the downside and narrowly miss 3σ on the upside.

The contrarian returns strategy has four days where returns exceed three standard deviations. These are May 7, November 9, and November 19, 2007, and January 23, 2008. There are none in August. Again, the confidence bands increase throughout July and reach their maximum width on August 6, before the hedge fund crisis is apparent.

Finally, the high minus low book to market portfolio is shown in figure 9.9. It has one day with a shock of 3σ magnitude and this is in the up direction on October 19. The confidence band increases well before the August hedge fund distress for this portfolio as well, but the interval continues to expand during the week of August 6–10.

The overall impression from this analysis is that, at least for these large-cap stocks, the model estimated through 2004 would be quite adequate for doing risk management for these types of dynamic long–short strategies. There is no evidence of 25σ events in these assets. It is also quite clear that there was evidence in the market before August that risks were increasing. In most cases, the risks were rising throughout July.

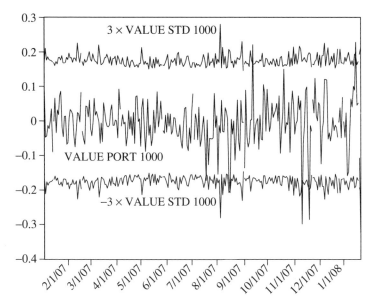

Figure 9.10. 3σ bands using 1,000-day historical variances and correlations with the value portfolio.

If volatilities and correlations are based on more naive methods that use stale information, then the outcome is not nearly so attractive. To illustrate this, consider the same value portfolio as in figure 9.7 but use a 1,000-day historical estimate of volatility and correlation. The resulting confidence interval is plotted in figure 9.10.

As we can see from this graph, there are now seven days of 3σ events rather than one. Furthermore, there are four days of 4σ events and one of a 5σ event. There are clearly many more 1σ and 2σ days than would normally be experienced. The use of a naive set of volatilities and correlations will lead to a misrepresentation of the risk, particularly when the economic environment is changing rapidly. Methods that update the correlations and volatilities as new information is received are likely to be much better measures of the risks to be experienced in the future.

This is a significant testing arena for the ability to anticipate correlations. Risk management is not only a critical feature of prudent asset management, but it is a key factor in choosing investment portfolios. The goal of making profits without bearing excessive risk can only be achieved with sophisticated methods of measuring and forecasting risk.

10
Credit Risk and Correlations

One of the most interesting developments in the financial marketplace over the last decade has been the growth in the volume and diversity of credit derivatives. These are contracts that provide insurance against defaults and therefore allow investors to manage their credit risk. The insurance can be purchased on either a single name or on baskets of securities. These allow investors to buy and sell risks associated with defaults either on a particular entity or on a portfolio. As financial markets have globalized, credit derivatives can now be found in portfolios all over the world. The ability to share risks allows investors to take only the risks they choose to take. For a discussion of this topic, see Hull (2006).

A credit default swap (CDS) provides insurance on a particular reference security. The purchaser pays a periodic coupon (usually quarterly) over the life of the CDS or until there is a default. In the event of a default, the purchaser can sell the defaulted bond of the reference entity for full face value to the CDS seller, or sometimes an equivalent cash settlement is stipulated. The seller receives the periodic premium until a default, at which time he must pay the loss on principal. The cash flows from selling protection are essentially the same as those from owning the reference bond, except that the principal is not exchanged. The size of the premium in CDS contracts is thus fundamentally linked to the probability of default and the recovery rate in case of default. The premium is often called a spread as it is the return above the riskless rate for bearing the default risk.

Models of default are sometimes formulated in terms of a structural model of equity prices and sometime in terms of reduced-form or direct-default predictions. The focus here will be on structural models that integrate the concepts developed thus far. Many excellent references discuss the two approaches: see, for example, Lando (2004), Schönbucher (2003), and McNeil et al. (2005). Longstaff and Rajan (2008) implement a reduced-form model on a data set very similar to the one used here. The structural model is based on Merton (1974) and reflects the fact that equity can be viewed as an option on the value of the firm. Default occurs when the price of equity goes to 0 and the bondholders own the remaining value of the firm. More realistically, a default will occur at a low but nonzero value of equity. Thus it is natural to think of the probability of default as the probability that equity values fall below some default frontier. If the probability of default is known, then the default frontier can be calculated based on models of future equity prices such as those developed in previous chapters. The market price of CDSs gives a market view of the probability of default. Approximately, the CDS spread divided by the recovery rate is often viewed as the risk-neutral probability of default. Various risk management systems calculate the default frontier from more fundamental considerations and measure the "distance to default" as the distance equity prices would fall at a default. From such an analysis one can calculate the probability of default and the price of a CDS. One can also calculate the risk of a CDS.

The important observation to take from this analysis is the close relationship between default probabilities and equity volatility. If the volatility of equity prices increases, then the probability of crossing any particular default frontier increases. If volatility is asymmetric, then the multiperiod distribution of returns will be negatively skewed and the probability of crossing the default frontier will be higher. Thus models of extreme moves in multiperiod equity returns are at the heart of models of default probabilities and CDS pricing in this structural context.

For investors, the risk of default might be much more manageable if defaults were all independent. In this case, a large portfolio with small investments in many bonds would never face much risk. However, when defaults are correlated, the probability of a large number of defaults occurring is substantial. The derivative contract that provides this insurance is generically described as a collateralized debt obligation (CDO). A CDO is a portfolio of assets of various forms that is sold to investors in various degrees of riskiness. This is accomplished by dividing the portfolio into tranches that differ in seniority. The first losses that occur are attributed entirely to the lowest, or equity, tranche. After a fixed fraction of the portfolio value is exhausted, the next tranches suffer the losses, and so on. After the equity tranche, the next lowest are the mezzanine or junior tranches, and the highest are the senior or super-senior tranches. The mezzanine tranches are more expensive to purchase and have a lower rate of return but also a lower default risk than the equity tranche. Finally, the senior tranches are the most expensive and receive the lowest return because they are nearly immune from defaults. In this way a portfolio that has some bad loans and some good loans that cannot be distinguished at the beginning can be sorted into high-grade and low-grade investments. Because ratings agencies typically, at least until the summer of 2007, rated the senior and super-senior tranches AAA regardless of the contents of the portfolio, the structure created both investment-grade and high-yield paper from any pool of loans. Das (2007) points out the importance of Basel II's reliance on ratings.

The appeal of this structure led to a booming business in credit markets. New types of loans were invented to fill these CDOs. We now see that the big business in subprime mortgages was fueled by this demand for assets to put into such portfolios. Similar explanations can be offered for the low-quality bonds that private equity was able to sell to finance its buyouts. Student loans, home equity loans, commercial loans, credit card balances, and many other asset types found their way into CDOs. An important type of CDO, called a synthetic CDO, contains CDSs on a collection of names. In this case the investor is selling credit protection and receives a periodic premium but is responsible for paying the losses associated with credit events. The variations on this structure are endless when it is realized that tranches of CDOs can be used as assets in other CDOs, leading to CDO squared and CDO cubed contracts. Monoline insurers provided protection for some of these instruments and many CDSs were written against CDO tranches. The investment-grade tranches of CDOs typically received a slightly higher return than comparable corporate debt, so many banks and other institutions leveraged these securities to achieve better returns. Important participants in this strategy included many of the originating brokers: for more on this, see UBS (2008) and IMF (2008), for example.

The bottom line is that there were enormous quantities of CDO tranches spread over the global financial markets. As many of these were funded by very risky debt, the investors—as well as the ratings agencies and even the originators themselves—must have believed that the senior tranches were essentially riskless regardless of their components. Time has proven that this was not correct. Massive losses have been realized in subprime mortgage CDOs and to a lesser extent in other types of CDOs.

Why were so many people surprised by these losses and how did they underestimate the risks? Two important changes that directly influenced these derivative securities were the changes in average default rates and the changes in default correlations. In the structural model of default, these are natural consequences of changes in equity volatility and equity correlations.

Consider the standardized synthetic CDO called CDX NA IG8, which was issued on March 20, 2007, and has a five-year maturity. It includes CDSs of 125 names and is equally weighted. The name includes "NA" for North America and "IG" for investment grade. The number 8 indicates

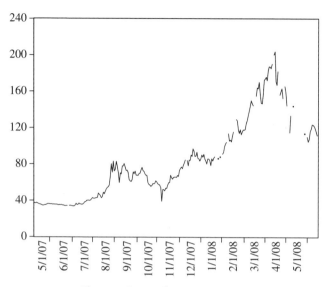

Figure 10.1. The CDX spread.

a particular synthetic CDO. An equally weighted index of these CDSs is also reported and tradeable; it is called CDX. Over the next year the spread for the index rose from under 40 basis points to more than 200 basis points. As these can roughly be considered as the spread between riskless rates and risky investment-grade debt, this is a dramatic increase (see figure 10.1). In the later period, there are some missing observations as these contracts were only sporadically traded.

Over the same period the volatility of the equity markets rose substantially as well. Using a TARCH model of the S&P 500 estimated through 2006, the daily update of the forecast standard deviation is plotted as a dashed line on the same graph but on a different scale in figure 10.2.

The relationship between equity volatilities and CDS default rates is apparent in these plots. Rising volatility is associated with rising CDS spreads and higher default rates. Thus one explanation for the failure to foresee the possibility that default rates might rise is that the low-volatility period lulled investors into thinking that volatility would not rise again. This is only partly credible because the term structure of volatility reflected in long-maturity option prices was rising very steeply. It is more likely that the link between equity volatility and default rates was not part of the risk management protocol.

In a similar way, the pricing of CDO tranches depends upon correlations. When default correlations are low, the risk of the senior tranches is negligible, but when correlations are high, the risk is almost the same as for the equity tranche. Clearly, when correlations are 1, then either

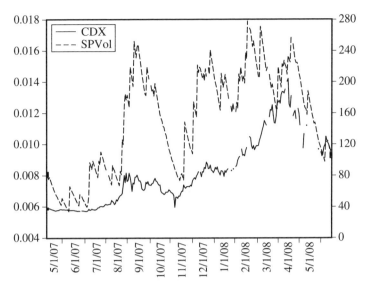

Figure 10.2. CDX and S&P 500 volatility.

everything defaults or nothing does, so the risks are the same for all tranches.

In figure 10.3, the midquotes of the bid and ask prices of tranche spreads for this CDO are shown over this turbulent year. The equity tranche for this CDO is the first 3% of the losses. Junior tranches are defined to take 3–7%, 7–10%, and 10–15% of the losses. The senior tranche absorbs the final 30–100% so it only pays losses when 30% of the assets default. It is clear that all the spreads increased dramatically. The 3–7% tranche rose from about 100 basis points to almost 1,000 basis points. At the same time the senior tranche from 30% to 100% rose from 1.5 basis points to more than 70 basis points. While the influence of the default rate is very apparent here, the effect of rising correlations can also be detected.

Plotting in figure 10.4 the three mezzanine spreads against the senior spread reveals that the relation is somewhat nonlinear. The more risky spreads did not increase in proportion to the spreads of the senior tranche. This effect is what would be expected from rising correlations.

These plots show, in a qualitative fashion, that the increase in spreads was even more dramatic in the senior tranches than in the junior tranches. This supports the view that rises in correlation, as well as those in volatility, are important in understanding the losses. By the same argument, it was the inability to foresee the possibility of rising volatilities and correlations that led so many investors and market participants to take on more risk than they would have chosen.

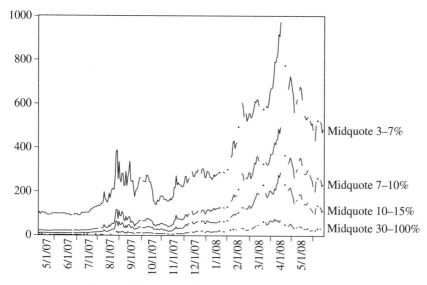

Figure 10.3. Tranche spreads of CDX NA IG8.

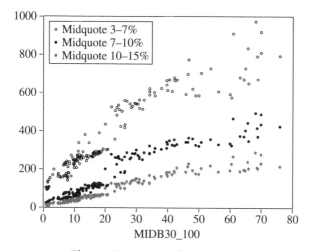

Figure 10.4. Tranche spreads.

It is important to assess more quantitatively the importance of volatility and correlation in explaining these changes in pricing. Taking default rates as given, how sensitive are the various tranches to correlations? A simple approach to this problem using the FACTOR ARCH model that has been discussed here was proposed in Berd et al. (2007). In this model, the market factor follows an asymmetric volatility process so that multiperiod returns have negative skewness. As a consequence, extreme declines in individual stocks are typically associated

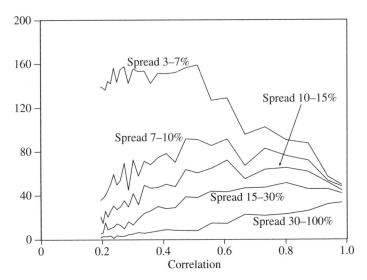

Figure 10.5.   Tranche spreads with correlation.

with extreme market declines. Defaults therefore often come in clusters. For some possible outcomes of the market factor there are no defaults, while for others there are many. The overall distribution depends on the relative importance of these two scenarios.

Using a model similar to that of Berd et al. and assuming a five-year horizon with a 60% recovery rate, the tranche spreads for the CDX can be computed with a range of assumptions. This model simulates the market return from an asymmetric GARCH model using historically filtered returns. The method for this simulation has recently been shown by Barone-Adesi et al. (2008) to be quite effective in pricing index options. In fact, the risk-neutral simulations run in that paper could be the same simulations used to price the CDX tranches. The noise in these figures is simply due to the randomness in the simulations.

In the first experiment, correlation is varied while everything else is held constant. The market return distribution is held constant but the idiosyncratic volatility is allowed to fall. As this parameter ranges from a daily standard deviation of 0.05 to 0.0018, the correlations rise from 0.2 to almost 1. As a result, the pricing of the various tranches changes markedly. The spreads of the high-risk tranches come down, while the spreads of the low-risk tranches rise. Ultimately, all the tranches are priced identically. This is a very extreme situation but it does reveal the important role of correlations.

Similarly, the model can take correlations as fixed and vary default probabilities. In this case, the prices of all tranches will increase and the

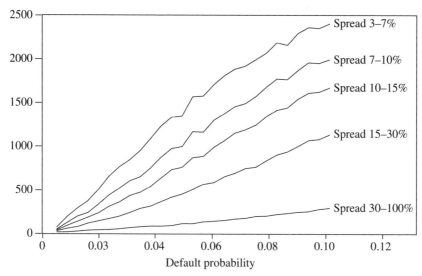

Figure 10.6. Tranche spreads by default probability.

relationship is roughly linear. This is shown in figure 10.6. As default probabilities go from 0.5% to 10%, the spread on the 3–7% tranche rises from 75 basis points to 2,400 basis points, or 24%. The spread for the senior tranche rises to 2.9%.

While this model makes some very strong assumptions, it reveals the important role of time-varying volatilities and correlations in pricing CDO tranches. The ability to incorporate the range of correlation models discussed in this book into the pricing of such complex derivative contracts is a topic of ongoing research.

Econometric Analysis of the DCC Model

In this chapter we turn to more rigorous econometric topics. The asymptotic properties of estimates of the DCC model are developed. The novel pieces are the analysis of correlation targeting and some alternatives, and the asymptotic distribution of DCC incorporating the two-step methodology and correlation targeting. Much of this material has previously been reported in Engle and Sheppard (2005b) and details of proofs will not be repeated here.

11.1 Variance Targeting

Most multivariate covariance models have a matrix of intercepts that must be estimated and which contains $\frac{1}{2}n(n-1)$ unknown parameters. For example, consider the simplest scalar multivariate GARCH model given by

$$H_t = \Omega + \alpha y_{t-1} y'_{t-1} + \beta H_{t-1}. \tag{11.1}$$

The most difficult part of the estimation is the intercept matrix. This is especially difficult if α and β sum to a number very close to 1, as the intercept matrix will be very small but must remain positive definite. A similar but less complicated situation occurs for a simple GARCH model in which the intercept must be positive but often very small.

It was proposed by Engle and Mezrich (1996) that the intercept matrix in such equations be estimated not by maximum likelihood but by an auxiliary estimator that is essentially a moment condition. The auxiliary estimator is simply given by

$$\hat{\Omega} = \bar{S}(1 - \alpha - \beta), \quad \bar{S} = \frac{1}{T} \sum_{t=1}^{T} y_t y'_t. \tag{11.2}$$

They called this "variance targeting," as it forced the variance–covariance matrix to take on a particular and plausible value. Such a moment condition is particularly attractive since it will be consistent regardless of whether the model (11.1) is correctly specified. To obtain an estimating equation for the remaining two parameters, (11.2) is substituted into (11.1) to give

$$H_t = \bar{S} + \alpha(y_{t-1}y'_{t-1} - \bar{S}) + \beta(H_{t-1} - \bar{S}). \tag{11.3}$$

This model can then be estimated by maximizing the likelihood functions with respect to the remaining parameters. The net result is that only 2, rather than $2 + \frac{1}{2}n(n-1)$, parameters need to be found from a nonlinear maximization. Of course, full MLE is an asymptotically efficient estimator and this two-step estimator is not. However, if the normality assumption is not satisfied, then both estimators are QMLE and the relative efficiency is not known for sure.

It remains to determine the standard errors of the estimated parameters of (11.3). Surely, conventional estimators that do not recognize the first-step estimation are likely to overestimate the precision of the estimates? We address this below.

11.2 Correlation Targeting

In the DCC model, correlation targeting is applied in the estimation of the mean-reverting model. It is, however, only an approximation as the estimating equation is nonlinear. Equation (4.36) shows that

$$Q_t = \Omega + \alpha\varepsilon_{t-1}\varepsilon'_{t-1} + \beta Q_{t-1}, \tag{11.4}$$

and averaging over all the time periods assuming large T,

$$\bar{Q} = \frac{1}{T}\sum_{t=1}^{T} Q_t \simeq \Omega + \alpha\bar{S} + \beta\bar{Q}, \quad \bar{S} = \frac{1}{T}\sum_{t=1}^{T}\varepsilon_t\varepsilon'_t, \tag{11.5}$$

where $\bar{S}$ is the sample correlation matrix of the standardized residuals, since the variables are all volatility adjusted and have a variance of 1. If we also assume that

$$\bar{Q} = \bar{S}, \tag{11.6}$$

then

$$\Omega = (1 - \alpha - \beta)\bar{S}. \tag{11.7}$$

This is called "correlation targeting" or perhaps "first-order correlation targeting."

It has frequently been observed, especially for large problems in real data or in simulations, that there is a downward bias in α. This possibly comes from the assumption made in (11.6).

A better approximation, which we might call "*second-order correlation targeting*," can be derived as follows. The computed correlation matrix R_t is given by

$$R_t = \{\text{diag } Q_t\}^{-1/2} Q_t \{\text{diag } Q_t\}^{-1/2}, \tag{11.8}$$

and its average can be defined as

$$\bar{R} = \frac{1}{T} \sum_{t=1}^{T} R_t. \tag{11.9}$$

Because of the nonlinearity in the definition of R, the mean of R is not the same as the mean of Q. We can show that the expected value of R is the expected unconditional correlation, $\bar{S}$:

$$E(\bar{S}_{i,j}) = \frac{1}{T} E \sum_{t=1}^{T} \varepsilon_{i,t}\varepsilon_{j,t} = \frac{1}{T} E \sum_{t=1}^{T} E_{t-1}(\varepsilon_{i,t}\varepsilon_{j,t}) = \frac{1}{T} E \sum_{t=1}^{T} R_{i,j,t} = E(\bar{R}_{i,j}). \tag{11.10}$$

As $T \to \infty$ the differences should converge in probability to 0, giving

$$\text{plim} \, |\bar{S} - \bar{R}| = 0. \tag{11.11}$$

We now seek a relation

$$\bar{R} = \bar{Q} + W. \tag{11.12}$$

For one correlation at time t this can be expressed as

$$R_{i,j,t} = \frac{Q_{i,j,t}}{\sqrt{Q_{i,i,t} Q_{j,j,t}}} = Q_{i,j,t} + \text{Taylor series}. \tag{11.13}$$

Expanding around the point ($Q_{i,i} = 1$, $Q_{j,j} = 1$) gives

$$\begin{aligned}
R_{i,j,t} &= Q_{i,j,t} - \tfrac{1}{2} Q_{i,j,t}[(Q_{i,i,t} - 1) + (Q_{j,j,t} - 1)] \\
&= Q_{i,j,t}[2 - \tfrac{1}{2}(Q_{i,i,t} + Q_{j,j,t})],
\end{aligned} \tag{11.14}$$

so that

$$\bar{R}_{i,j} = \bar{Q}_{i,j} - \frac{1}{T} \sum_{t=1}^{T} Q_{i,j,t}(\tfrac{1}{2}(Q_{i,i,t} + Q_{j,j,t}) - 1). \tag{11.15}$$

Hence the correction depends negatively upon the covariance between diagonal and off-diagonal elements over time. There is no correction for the diagonal elements. This correction term is simply the average covariance between the diagonal and off-diagonal elements of Q:

$$W_{i,j} = -0.5(\text{Cov}(Q_{i,i,t}, Q_{i,j,t}) + \text{Cov}(Q_{j,j,t}, Q_{i,j,t})). \tag{11.16}$$

Gathering this into a matrix we get the expression (11.12). Substituting (11.11) and (11.12) into (11.5) eliminates the unobservables $\bar{Q}$ and $\bar{R}$, giving the second-order correlation target

$$\Omega = \bar{S}(1 - \alpha - \beta) - W(1 - \beta). \tag{11.17}$$

This adjustment will make numerical evaluation of the likelihood function slightly slower and will change numerical derivatives. However, it will hopefully provide an improvement in the accuracy of forecasts. Clearly, the same criterion can be evaluated more accurately further expansion of the Taylor series. The guarantee of positive-definite correlation matrices could, however, be compromised.

Finally, an alternative to approximate correlation targeting lies in reformulating the DCC model to make it exact. This was done in one way by Tse and Tsui (2002) and in another way by Aielli (2006). The method of Aielli is a neat solution. The DCC updating in (11.4) is replaced by

$$Q_t = \Omega + \alpha \operatorname{diag}(Q_{t-1})^{1/2} \varepsilon_{t-1} \varepsilon_{t-1}' \operatorname{diag}(Q_{t-1})^{1/2} + \beta Q_{t-1}. \tag{11.18}$$

Correlations are still given by (11.8), so

$$\operatorname{diag}(Q_t)^{1/2} E_{t-1}(\varepsilon_t \varepsilon_t') \operatorname{diag}(Q_t)^{1/2} = \operatorname{diag}(Q_t)^{1/2} R_t \operatorname{diag}(Q_t)^{1/2} = Q_t. \tag{11.19}$$

Therefore, taking unconditional expectations of both sides of (11.18) gives the difference equation, which can easily be solved to give an unconditional expression and multistep forecasts for Q:

$$\begin{aligned} E(Q_t) &= \Omega + (\alpha + \beta) E(Q_{t-1}) \\ &\Rightarrow E(Q_t) = \Omega / (1 - \alpha - \beta) \\ &\Rightarrow E_t(Q_{t+k}) = E(Q) + (\alpha + \beta)(E_t(Q_{t+k-1}) - E(Q)). \end{aligned} \tag{11.20}$$

In this model, correlation targeting is not as simple as before but it is exact, at least for large T. Define

$$\bar{S}^* = \frac{1}{T} \sum_{t=1}^{T} \operatorname{diag}(Q_t)^{1/2} \varepsilon_t \varepsilon_t' \operatorname{diag}(Q_t)^{1/2}. \tag{11.21}$$

Taking unconditional expectations of (11.21) using (11.19) and (11.20) gives the moment condition

$$E(\bar{S}^*) = \Omega / (1 - \alpha - \beta), \tag{11.22}$$

which replaces (11.7) and yields

$$Q_t = \bar{S}^* + \alpha(\operatorname{diag}(Q_{t-1})^{1/2} \varepsilon_{t-1} \varepsilon_{t-1}' \operatorname{diag}(Q_{t-1})^{1/2} - \bar{S}^*) + \beta(Q_{t-1} - \bar{S}^*). \tag{11.23}$$

To estimate this model we must iterate between maximizing the likelihood based on (11.23) and estimating the adjusted covariance of returns in (11.22).

It is not clear how effective these two adjustments to DCC will be, but there is reason to believe that they could be helpful. In addition, these effects become more important when asymmetric models are used or exogenous variables are introduced into the Q equations. Future research will have to clarify these issues.

11.3 Asymptotic Distribution of DCC

Engle and Sheppard (2005b) establish the asymptotic distribution of the DCC estimator taking into account both the two-step estimation method, which first estimates variances and then estimates correlations, and the use of correlation targeting. In fact, it is simplest to think of this as merely a three-step method: first variances, second correlations, third the DCC parameters. It is a QMLE theorem in the sense that it does not assume that the underlying conditional distribution is a multivariate normal. The distribution could be some other joint distribution with fat tails in some or all directions but with a covariance structure that is correctly given by the DCC process. An interesting way to think of the true model is in terms of the meta-Gaussian or meta Student t copulas described in chapter 2. In this case, each marginal distribution can be different and can have different tail behavior, but the copula is a Gaussian or a Student t copula which measures the dependence among contemporaneous shocks through their correlation.

Theorems 1 and 2 in Engle and Sheppard (2005b) establish the consistency and asymptotic normality of the DCC model under a series of assumptions. The theorems are essentially GMM theorems without overidentifying moment conditions as this is a simple way to prove results for multistep estimators. Many of these assumptions are regularity conditions designed to ensure that moments are continuous and approach well-defined limits.

Let me sketch the setup and results. The theorem itself needs no new proof as it is in several books (see, for example, Newey and McFadden 1994; Wooldridge 1994; White 1984). Let the DCC model be given by

$$
\left.
\begin{aligned}
y_t | \mathbb{F}_{t-1} &\sim N(0, D_t R_t D_t), \\
D_{i,i,t}^2 &= \theta_{i,0} + \theta_{i,1} y_{i,t-1}^2 + \theta_{i,3} D_{i,i,t-1}^2, \\
R_t &= \text{diag}\{Q_t\}^{-1/2} Q_t \, \text{diag}\{Q_t\}^{-1/2}, \\
Q_t &= \Psi + \phi_1(\varepsilon_{t-1}\varepsilon_{t-1}' - \Psi) + \phi_2(Q_{t-1} - \Psi).
\end{aligned}
\right\}
\tag{11.24}
$$

There are three sets of parameters: $\xi = (\theta, \Psi, \phi)$, which corresponds to the variance models, the correlations, and the DCC. The log likelihood depends on all of these, but in a rather special way. The likelihood for observation t can be separated out from (4.34) as

$$L_t(\theta, \Psi, \phi; y_t) = L_{1,t}(\theta; y_t) + L_{2,t}(\theta, \Psi, \phi; y_t). \tag{11.25}$$

The first set of moment conditions is given by

$$g_{1T}(y, \theta) = \frac{1}{T} \sum_{t=1}^{T} \nabla_\theta L_{1,t}. \tag{11.26}$$

The second set of moment conditions can best be expressed in terms of the standardized returns from the first stage. These depend implicitly on the vector of parameters θ. The moments are the unique elements of the following expression:

$$g_{2T}(\theta, \Psi) = \frac{1}{T} \sum_{t=1}^{T} (\Psi - \varepsilon_t \varepsilon_t'). \tag{11.27}$$

The third set of moment conditions is given by

$$g_{3T}(\theta, \Psi, \phi) = \frac{1}{T} \sum_{t=1}^{T} \nabla_\phi L_{2,t}(\theta, \Psi, \phi; y_t). \tag{11.28}$$

There is one moment condition for each parameter, hence there will in most cases be a set of parameters that makes the moment conditions (11.26), (11.27), and (11.28) all equal to zero. From the structure of these conditions, θ must set (11.26) to zero. Using this value of θ, Ψ sets (11.27) to zero. Finally, the solution to (11.28) takes the solution values of the first two sets of parameters as given and solves for ϕ. This recursive structure of the moment conditions makes the estimator a sequential three-step procedure.

Theorem 1 of Engle and Sheppard (2005b) states that under a collection of regularity conditions, the estimator is consistent:

$$\hat{\xi}_T \xrightarrow{\text{p}} \xi_0. \tag{11.29}$$

Theorem 2 states that under an additional set of regularity conditions, the estimator is asymptotically normal with a covariance matrix of the familiar sandwich form:

$$\sqrt{T}(\hat{\xi} - \xi_0) \xrightarrow{\text{d}} N(0, A_0^{-1} B_0 A_0^{-1}), \tag{11.30}$$

where A_0 and B_0 are given by

$$A_0 = E \left[\begin{bmatrix} \nabla_\theta g_{1T}(\theta_0) & 0 & 0 \\ \nabla_\theta g_{2T}(\theta_0, \Psi_0) & \nabla_\Psi g_{2T}(\theta_0, \Psi_0) & 0 \\ \nabla_\theta g_{3T}(\theta_0, \Psi_0, \phi_0) & \nabla_\Psi g_{3T}(\theta_0, \Psi_0, \phi_0) & \nabla_\phi g_{3T}(\theta_0, \Psi_0, \phi_0) \end{bmatrix} \right],$$

$$B_0 = \text{AVAR} \left\{ \sum_{t=1}^T \begin{pmatrix} g_{1t}(\xi_0) & g_{2T}(\xi_0) & g_{3t}(\xi_0) \end{pmatrix} \right\}.$$

(11.31)

This result shows how to estimate covariance matrices for DCC models correcting for the multistage estimation strategies. Although this is implemented in Engle and Sheppard (2005b), it is not widely used. The derivatives in (11.31) can be calculated numerically but, fortunately, some are very simple. For example, the largest block, which is $\nabla_\Psi g_{2T}(\theta_0, \Psi_0)$, is simply the identity matrix. It is likely that changes in the volatility parameters will have relatively little effect on the correlation estimates, and therefore the second and third blocks in the first column may be small. Thus the effect of the Hessian on the standard errors may not be big. The estimation of B is generally done by summing the squares and cross products of the moment conditions and correcting for autocorrelation if necessary. If the model is correctly specified, these moments should be martingale difference processes and thus a simple estimator such as

$$\hat{B} = \frac{1}{T} \sum_{t=1}^T \begin{pmatrix} g_{1t} g'_{1t} & g_{1t} g'_{2t} & g_{1t} g'_{3t} \\ g_{2t} g'_{1t} & g_{2t} g'_{2t} & g_{2t} g'_{3t} \\ g_{3t} g'_{1t} & g_{3t} g'_{2t} & g_{3t} g'_{3t} \end{pmatrix}$$

(11.32)

may be adequate. For misspecified models, various serial correlation-consistent covariance estimators could be substituted for (11.32). The performance of these estimators is a topic for future research.

12
Conclusions

Financial decision making and risk analysis requires accurate and timely estimates of correlations today and how they are likely to evolve in the future. New methods for analyzing correlations have been developed in this book and compared with existing methods. The DCC model and its extensions, such as the FACTOR DCC model, are promising tools. These models are simple and powerful and have been shown to provide good assessment of risk and temporal stability in the face of dramatic shifts in the investment environment.

We began by looking at the typical pattern of correlations across asset classes and countries. The economic analysis of correlations focused on the role of news in moving the prices of assets, whereby the correlation of news events across assets was a central determinant of the correlation of returns. As a consequence, the changing intensity and importance of various news sources impacts both volatilities and correlations across assets. It is unlikely that traditional factor models will be adequate to model such dynamics.

In the second chapter, the concept of conditional correlations is introduced to formulate the problem of modeling future correlations conditional on past information. More general measures of dependence based on copulas are developed and these provide a framework for asking about extremes and other types of nonnormal behavior. Extreme returns are particularly relevant for credit risk measures focused on defaults. From a portfolio construction point of view, accurate correlations allow investors to better optimize between risk and return and provide a laboratory in which to test the precision of correlation models.

Time-varying correlation models are introduced in chapter 3: from historical correlation models through multivariate GARCH to orthogonal GARCH and, finally, to dynamic conditional correlation, the model that takes a central role in the book. The DCC model has three steps, which are called DE-GARCHING, quasi-correlation estimation, and re-scaling. These are then examined in some detail with a discussion of the overall estimation method. Succeeding chapters then examine the performance of the simple DCC model in both the simulation and empirical settings.

Correlations change over time in response to a variety of economic events. The correlation between stocks and bonds is examined first. For much of the sample these asset classes are positively correlated; however, when the economy is overheated, good news for the stock market becomes bad news for the bond market and the correlations become seriously negative. Correlations between large-cap stocks change over time when the companies change their lines of business or when latent factors such as energy prices or natural factors such as the market as a whole become more volatile. This leads to interesting variations in correlations and betas over time. An examination of the comovement of stocks and bonds in a global setting finds strong evidence of asymmetric effects in the stock volatilities and correlations. In Cappiello et al. (2007), the DCC model is generalized to allow country-specific coefficients in the correlation process. This GDCC model illustrates important flexibilities in the DCC methodology. In this data set, correlations among European equity and bond returns show a substantial increase around the formation of the euro; however, these correlations also arise between countries that did not adopt the currency.

To extend this analysis to systems of very large dimension, computational innovations are required. A new estimation method called the MacGyver method is introduced. This method no longer depends on estimating the whole system of correlations at the same time. Instead it estimates all the bivariate models and then extracts the best estimates of the unknown parameters from these separate problems. The simulation performance is very attractive and the empirical applications are very sensible. The limit on the size of DCC models that can be estimated is now very large.

In typical finance applications, large-dimension systems are modeled with factor structures. In these models, a small number of factors are used to approximate a large number of assets. In creating factor versions of the correlation models based on DCC, several very useful new models are introduced. The standard one-factor model underlying the CAPM that was introduced by Engle et al. (1990b) is extended to allow idiosyncratic volatilities to change over time and to allow the residuals to have some dynamic correlation. The final model is called the FACTOR DCC and has the ability to model correlations when the number of factors and their volatility are varying and when the factor loadings are changing over time. This great flexibility comes with very few new parameters and with estimation that is hardly more complex than the DCC itself.

The FACTOR DCC is an attractive theoretical model but it is important to see how it performs. Using eighteen large-cap stocks from 1994 through 2004, the models are all estimated and compared. Portfolios were optimized using each of these models and the FACTOR DCC and DCC models were found to be the best two approaches. The constant, 100-day historical, and FACTOR ARCH models produced inferior portfolios. Although the differences in portfolio volatilities were rather small, they were systematic in the sense that the same models performed best for most of the cases.

The same experiment was repeated for a post-sample data set that included the low-volatility period 2005–6 and the high-volatility period starting in August 2007. Using the parameters estimated through 2004, the volatilities and correlations were updated daily. The computed market volatility rose dramatically from August through October and then again from November through January. Correspondingly, correlations rose over this period, reaching levels above 50% in January for the FACTOR DCC. These values are comparable with the highest of 2001–3. As a consequence, many portfolio strategies were assuming substantially more risk than they had been in the low-volatility period from 2003 to 2006. Furthermore, the rise in correlations naturally affected the risk and pricing of correlation-sensitive derivatives such as CDOs. It remains to investigate whether changing equity correlations can be associated with changing housing correlations, which would then be a key point in understanding the increase in risk of the senior tranches of subprime backed securities.

The hedging experiment was repeated for the forecast data set and the results were quite similar, with the DCC model now proving slightly more accurate than the FACTOR DCC model. The differences between these models are not as systematic as before. Both models easily dominated the constant, 100-day historical, and FACTOR ARCH models. This is an impressive performance as it means that even using parameter values that are several years out of date, the model still produces good portfolios.

These same volatilities and correlations can be used to measure the risk of any portfolio. In particular, Khandani and Lo (2007) and many newspaper articles claim that the market in August 2007 behaved in a very unusual fashion as long-/short-value portfolios lost money because the long prices declined while the short prices went up. There were claims that multiple 25σ events occurred. To determine whether risk management using the DCC model would have recognized this effect, long-/short-value portfolios were constructed using a variety of signals. The volatility forecasts for each portfolio were constructed. In each case, the volatilities of these portfolios began rising in July and were already high by August 2007. There were no 3σ days in August for any of the portfolios, although such events did occur in November 2007 and in January 2008. No 4σ events were observed.

The combination of the theoretical attractions of these models with their empirical performance and their statistical simplicity provides a good argument for applying the FACTOR DCC model or some variation of it in everyday practical risk management. There does not appear to be any obstacle to extending this to multiple asset classes, multiple factors, and high-dimensional systems.

Such analyses are then extended in order to value correlation-sensitive derivatives, such as CDOs and other asset-backed securities, and to measure their risks. Similarly, the trading and hedging of baskets of options and dispersion trades can be modeled within this framework. The financial sector experienced extraordinary losses on portfolios of derivatives based on subprime mortgages and other assets. These products are sensitive to correlations and volatilities. From examining prices of standardized synthetic CDOs, the sensitivity to these parameters is investigated and it is found that substantial risk is indeed associated with changes in these parameters.

The financial marketplace is an infinitely complex, continuously evolving process. The models developed in this book have the potential to adapt to unforeseen changes in the financial environment and hence give a dynamic picture of correlations and volatilities. These methods are naturally short-run methods focusing on what can happen in the near future. Risk management must necessarily also be concerned with the longer run. Conveniently, the factor versions of these models allow us to model the determinants of factor volatilities, which are the primary determinants of longer-term volatilities and correlations.

This book is offered with the hope that with this type of sound modeling, investors will one day be offered the chance to take only the risks they want to take and leave the rest for others with different appetites. Indeed, the underpinning of our theories of finance is the choice between risk and return for all investors, and this requires ever-changing assessments of risk in a world of such great and interesting complexity.

References

Aielli, G. P. 2006. Consistent estimation of large scale dynamic conditional correlations. Working Paper, Department of Statistics, University of Florence.

Alexander, C. 2002. Principal component models for generating large GARCH covariance matrices. *Economic Notes* 23:337-59.

Alexander, C., and A. Barbosa. 2008. Hedging index exchange traded funds. *Journal of Banking & Finance* 32:326-337.

Ammer, J., and J. Mei. 1996. Measuring international economic linkages with stock market data. *Journal of Finance* 51:1743-63.

Andersen, T. G., T. Bollerslev, P. F. Christoffersen, and F. X. Diebold. 2005. Practical volatility and correlation modeling for financial market risk management. Working Paper 1-39, Financial Institutions Center at the Wharton School.

Anderson, H. M., and F. Vahid. 2007. Forecasting the volatility of Australian stock returns: do common factors help? *Journal of Business & Economic Statistics* 25:76-90.

Anderssona, M., E. Krylovab, and S. Vahamaa. 2008. Why does the correlation between stock and bond returns vary over time? *Applied Financial Economics* 18:139-51.

Ang, A., and J. Chen. 2002. Asymmetric correlations of equity portfolios. *Journal of Financial Economics* 63:443-94.

Asai, M., M. McAleer, and J. Yu. 2006. Multivariate stochastic volatility: a review. *Econometric Reviews* 25:145-75.

Audrino, F., and G. Barone-Adesi. 2006. Average conditional correlation and tree structures for multivariate GARCH models. *Journal of Forecasting* 25:579-600.

Baillie, R. T., and R. J. Myers. 1991. Bivariate GARCH estimation of the optimal commodity futures hedge. *Journal of Applied Econometrics* 6:109-24.

Barone-Adesi, G., R. F. Engle, and L. Mancini. 2008. A GARCH option pricing model with filtered historical simulation. *Review of Financial Studies* 21:1223-58.

Bartram, S. M., S. J. Taylor, and Y.-H. Wang. 2007. The euro and European financial market dependence. *Journal of Banking & Finance* 31:1461-81.

Bautista, C. C. 2006. The exchange rate-interest differential relationship in six East Asian countries. *Economics Letters* 92:137-42.

Bauwens, L., and S. Laurent. 2005. A new class of multivariate skew densities, with application to generalized autoregressive conditional heteroscedasticity models. *Journal of Business & Economic Statistics* 23:346-54.

Bauwens, L., and J. V. K. Rombouts. 2007. Bayesian clustering of many GARCH models. *Econometric Reviews* 26:365-86.

Bauwens, L., S. Laurent, and J. V. K. Rombouts. 2006. Multivariate GARCH models: a survey. *Journal of Applied Econometrics* 21:79-109.

Bekaert, G., and C. R. Harvey. 1995. Time-varying world market integration. *Journal of Finance* 50:403-44.

Bekaert, G., and G. Wu. 2000. Asymmetric volatility and risk in equity markets. *Review of Financial Studies* 13:1–42.

Berd, A., R. F. Engle, and A. Voronov. 2007. The underlying dynamics of credit correlations. *Journal of Credit Risk* 3:27–62.

Billio, M., and M. Caporin. 2005. Multivariate Markov switching dynamic conditional correlation GARCH representations for contagion analysis. *Statistical Methods & Applications* 14:145–61.

Billio, M., M. Caporin, and M. Gobbo. 2006. Flexible dynamic conditional correlation multivariate GARCH models for asset allocation. *Applied Financial Economics Letters* 2:123–30.

Bodurtha, J. N., and N. C. Mark. 1991. Testing the CAPM with time-varying risks and returns. *Journal of Finance* 46:1485–505.

Bollerslev, T. 1986. Generalized autoregressive conditional heteroskedasticity. *Journal of Econometrics* 31:307–27.

——. 1987. A conditionally heteroskedastic time series model for speculative prices and rates of return. *Review of Economics and Statistics* 69:542–47.

——. 1990. Modelling the coherence in short-run nominal exchange rates: a multivariate generalized ARCH model. *Review of Economics and Statistics* 72: 498–505.

——. Forthcoming. Glossary to ARCH (GARCH). In *Volatility and Time Series Econometrics: Essays in Honor of Robert F. Engle* (ed. T. Bollerslev, J. Russell, and M. Watson). Oxford University Press.

Bollerslev, T., and J. M. Wooldridge. 1992. Quasi-maximum likelihood estimation and inference in dynamic models with time-varying covariances. *Econometric Reviews* 11:143–72.

Bollerslev, T., R. F. Engle, and J. M. Wooldridge. 1988. A capital-asset pricing model with time-varying covariances. *Journal of Political Economy* 96:116–31.

Bollerslev, T., R. Y. Chou, and K. F. Kroner. 1992. ARCH modeling in finance: a review of the theory and empirical evidence. *Journal of Econometrics* 52:5–59.

Bollerslev, T., R. F. Engle, and D. Nelson. 1994. ARCH models. In *Handbook of Econometrics* (ed. R. F. Engle and D. McFadden), volume IV, pp. 2959–3038. Amsterdam: North-Holland.

Braun, P. A., D. B. Nelson, and A. M. Sunier. 1995. Good-news, bad-news, volatility, and betas. *Journal of Finance* 50:1575–603.

Burns, P., R. F. Engle, and J. Mezrich. 1998. Correlations and volatilities of asynchronous data. *Journal of Derivatives* (Summer):1–12.

Bystrom, H. N. E. 2004. Orthogonal GARCH and covariance matrix forecasting: the Nordic stock markets during the Asian financial crisis 1997–1998. *European Journal of Finance* 10:44–67.

Cajigas, J.-P., and G. Urga. 2006. Dynamic conditional correlation models with asymmetric multivariate Laplace innovations. Working Paper, Centre for Econometric Analysis, Cass Business School.

Campbell, J. H., and J. Ammer. 1993. What moves the stock and bond markets? A variance decomposition for long-term asset returns. *Journal of Finance* 48: 3–37.

Campbell, J. H., M. Lettau, B. Malkiel, and Y. Xu. 2001. Have individual stocks become more volatile? An empirical exploration of idiosyncratic risk. *Journal of Finance* 56:1–43.

Campbell, J. Y. 1991. A variance decomposition for stock returns. *Economic Journal* 101:157–79.

Campbell, J. Y., and L. Hentschel. 1992. No news is good news: an asymmetric model of changing volatility in stock returns. *Journal of Financial Economics* 31:281–318.

Campbell, J. Y., and R. J. Shiller. 1988a. The dividend–price ratio and expectations of future dividends and discount factors. *Review of Financial Studies* 1: 195–228.

———. 1988b. Stock prices, earnings, and expected dividends. *Journal of Finance* 43:661–76.

Caporin, M. 2007. Variance (non) causality in multivariate GARCH. *Econometric Reviews* 26:1–24.

Cappiello, L., R. F. Engle, and K. Sheppard. 2007. Asymmetric dynamics in the correlations of global equity and bond returns. *Journal of Financial Econometrics* 4:537–72.

Chan, F., C. Lim, and M. McAleer. 2005a. Modeling multivariate international tourism demand and volatility. *Tourism Management* 26:459–71.

Chan, F., D. Marinova, and M. McAleer. 2005b. Rolling regressions and conditional correlations of foreign patents in the USA. *Environmental Modelling & Software* 20:1413–22.

Chandra, M. 2005. Estimating and explaining extreme comovements in Asia-Pacific equity markets. *Review of Pacific Basin Financial Markets & Policies* 8: 53–79.

———. 2006. The day-of-the-week effect in conditional correlation. *Review of Quantitative Finance & Accounting* 27:297–310.

Chen, X., and Y. Fan. 2006. Estimation and model selection of semiparametric copula-based multivariate dynamic models under copula misspecification. *Journal of Econometrics* 135:125–54.

Chen, Y.-T. 2007. Moment-based copula tests for financial returns. *Journal of Business & Economic Statistics* 25:377–97.

Chiang, T. C., B. N. Jeon, and H. Li. 2007. Dynamic correlation analysis of financial contagion: evidence from Asian markets. *Journal of International Money & Finance* 26:1206–28.

Chib, S., F. Nardari, and N. Shephard. 2006. Analysis of high dimensional multivariate stochastic volatility models. *Journal of Econometrics* 134:341–71.

Chong, J. 2005. The forecasting abilities of implied and econometric variance-covariance models across financial measures. *Journal of Economics & Business* 57:463–90.

Chou, R. Y. 2005. Forecasting financial volatilities with extreme values: the conditional autoregressive range (CARR) model. *Journal of Money, Credit & Banking* 37:561–82.

Christodoulakis, G. A. 2007. Common volatility and correlation clustering in asset returns. *European Journal of Operational Research* 182:1263–84.

Ciner, C. 2006. A further look at linkages between NAFTA equity markets. *Quarterly Review of Economics & Finance* 46:338–52.

Colacito, R., R. F. Engle, and E. Ghysels. 2008. A component model for dynamic correlations. Manuscript, University of North Carolina.

Crespo Cuaresma, J., and C. Wojcik. 2006. Measuring monetary independence: evidence from a group of new EU member countries. *Journal of Comparative Economics* 34:24–43.

Das, S. R. 2007. Basel II: correlation related issues. *Journal of Financial Services Research* 32:17–38.

de Goeij, P., and W. Marquering. 2004. Modeling the conditional covariance between stock and bond returns: a multivariate GARCH approach. *Journal of Financial Econometrics* 2:531–564.

——. 2006. Macroeconomic announcements and asymmetric volatility in bond returns. *Journal of Banking & Finance* 30:2659–80.

Dellaportas, P., and I. D. Vrontos. 2007. Modelling volatility asymmetries: a Bayesian analysis of a class of tree structured multivariate GARCH models. *Econometrics Journal* 10:503–20.

Diebold, F. X., and R. S. Mariano. 2002. Comparing predictive accuracy. *Journal of Business & Economic Statistics* 20:253–63.

Diebold, F. X., T. A. Gunther, and A. S. Tay. 1998. Evaluating density forecasts with applications to financial risk management. *International Economic Review* 39:863–83.

Diebold, F. X., J. Hahn, and A. S. Tay. 1999. Multivariate density forecast evaluation and calibration in financial risk management: high. *Review of Economics & Statistics* 81:661–73.

Ding, Z., and R. F. Engle. 2001. Large scale conditional covariance matrix modeling, estimation and testing. *Academia Economic Papers* 29:157–84.

Duffee, G. R. 2005. Time variation in the covariance between stock returns and consumption growth. *Journal of Finance* 60:1673–712.

Eklund, B., and T. Terasvirta. 2007. Testing constancy of the error covariance matrix in vector models. *Journal of Econometrics* 140:753–80.

Engle, R. F. 1982. Autoregressive conditional heteroskedasticity with estimates of the variance of UK inflation. *Econometrica* 50:987–1008.

Engle, R. F. 1995. Introduction. In *ARCH: Selected Readings*, pp. xi–xviii. Advanced Texts in Econometrics. Oxford University Press.

——. 2001. GARCH 101: the use of ARCH/GARCH models in applied econometrics. *Journal of Economic Perspectives* Fall, V15N4.

——. 2002a. Dynamic conditional correlation: a simple class of multivariate generalized autoregressive conditional heteroskedasticity models. *Journal of Business & Economic Statistics* 20:339–50.

——. 2002b. New frontiers for ARCH models. *Journal of Applied Econometrics* 17:425–46.

——. 2004. Risk and volatility: econometric models and financial practice. *American Economic Review* 94:405–20.

Engle, R. F., and R. Colacito. 2006. Testing and valuing dynamic correlations for asset allocation. *Journal of Business & Economic Statistics* 24:238–53.

Engle, R. F., and S. Kozicki. 1993. Testing for common features. *Journal of Business & Economic Statistics* 11:369–80.

Engle, R. F., and K. F. Kroner. 1995. Multivariate simultaneous generalized ARCH. *Econometric Theory* 11:122–50.

Engle, R. F., and J. Marcucci. 2006. A long-run pure variance common features model for the common volatilities of the Dow Jones. *Journal of Econometrics* 132:7–42.

Engle, R. F., and J. Mezrich. 1996. GARCH for groups. *Risk* 9:36–40.

Engle, R. F., and V. K. Ng. 1993. Measuring and testing the impact of news on volatility. *Journal of Finance* 48:1749–78.

Engle, R. F., and A. J. Patton. 2001. What good is a volatility model? *Quantitative Finance* 1:237–45.

Engle, R. F., and J. G. Rangel. 2008. The Spline-GARCH model for low-frequency volatility and its global macroeconomic causes. *Review of Financial Studies* 21(3):1187–222.

Engle, R. F., and K. Sheppard. 2005a. Evaluating the specification of covariance models for large portfolios. Working Paper, University of California, San Diego.

——. 2005b. Theoretical properties of dynamic conditional correlation multivariate GARCH. Working Paper, University of California, San Diego.

Engle, R. F., and R. Susmel. 1993. Common volatility in international equity markets. *Journal of Business and Economic Statistics* 11:167–76.

Engle, R. F., D. F. Hendry, and J. F. Richard. 1983. Exogeneity. *Econometrica* 51: 277–304.

Engle, R. F., D. Kraft, and C. W. J. Granger. 1984. Combining competing forecasts of inflation based on a multivariate ARCH model. *Journal of Economic Dynamics and Control* 8:151–65.

Engle, R. F., T. Ito, and W. L. Lin. 1990a. Meteor showers or heat waves? Heteroskedastic intradaily volatility in the foreign-exchange market. *Econometrica* 58:525–42.

Engle, R. F., V. K. Ng, and M. Rothschild. 1990b. Asset pricing with a factor-ARCH covariance structure: empirical estimates for treasury bills. *Journal of Econometrics* 45:213–37.

Engle, R. F., E. Ghysels, and B. Sohn. 2008a. On the economic source of stock market volatility. Manuscript, University of North Carolina.

Engle, R. F., N. Shephard, and K. Sheppard. 2008b. Fitting and testing vast dimensional time-varying covariance models.

Epps, T. W. 1979. Comovements in stock prices in the very short run. *Journal of the American Statistical Association* 74:291–298.

Fernandes, M., B. de Sa Mota, and G. Rocha. 2005. A multivariate conditional autoregressive range model. *Economics Letters* 86:435–40.

Ferreira, M. A., and J. A. Lopez. 2005. Evaluating interest rate covariance models within a Value-at-Risk framework. *Journal of Financial Econometrics* 3:126–68.

Flavin, T. 2006. How risk averse are fund managers? Evidence from Irish mutual funds. *Applied Financial Economics* 16:1355–63.

French, K. R., G. W. Schwert, and R. F. Stambaugh. 1987. Expected stock returns and volatility. *Journal of Financial Economics* 19:3–29.

Gallo, G. M., and E. Otranto. 2007. Volatility transmission across markets: a multichain Markov switching model. *Applied Financial Economics* 17:659–70.

Giamouridis, D., and I. D. Vrontos. 2007. Hedge fund portfolio construction: a comparison of static and dynamic approaches. *Journal of Banking & Finance* 31:199–217.

Glosten, L. R., R. Jagannathan, and D. E. Runkle. 1993. On the relation between the expected value and the volatility of the nominal excess return on stocks. *Journal of Finance* 48:1779–801.

Goetzmann, W. N., L. Li, and K. G. Rouwenhorst. 2005. Long-term global market correlations. *Journal of Business* 78:1–38.

Guirguis, H., and R. Vogel. 2006. Asymmetry in regional real house prices. *Journal of Real Estate Portfolio Management* 12:293–98.

Hafner, C. M., and P. H. Franses. 2003. A generalized dynamic conditional correlation model for many assets. Econometric Institute Report, Erasmus University Rotterdam.

Hafner, C. M., D. van Dijk, and P. H. Franses. 2006. Semi-parametric modelling of correlation dynamics. In *Econometric Analysis of Financial and Economic Time Series* (ed. D. Terrell and T. B. Fomby), part A. Advances in Econometrics, volume 20, pp. 59-103. Elsevier.

Harris, R. D. F., E. Stoja, and J. Tucker. 2007. A simplified approach to modeling the co-movement of asset returns. *Journal of Futures Markets* 27:575-98.

He, C., and T. Terasvirta. 2004. An extended constant conditional correlation GARCH model and its fourth-moment structure. *Econometric Theory* 20:904-26.

Hull, J. 2006. *Options, Futures and Other Derivatives*, 6th edn. Upper Saddle River, NJ: Prentice Hall.

IMF. 2008. *Global Financial Stability Report: Containing Systemic Risks and Restoring Financial Soundness*. Washington, DC: International Monetary Fund.

Joe, H. 1997. *Multivariate Models and Dependence Concepts*. London: Chapman & Hall.

Jondeau, E., and M. Rockinger. 2006. The copula-GARCH model of conditional dependencies: an international stock market application. *Journal of International Money & Finance* 25:827-53.

Kawakatsu, H. 2006. Matrix exponential GARCH. *Journal of Econometrics* 134:95-128.

Kearney, C., and A. J. Patton. 2000. Multivariate GARCH modeling of exchange rate volatility transmission in the European monetary system. *Financial Review* 35:29-48.

Kearney, C., and V. Poti. 2004. Idiosyncratic risk, market risk and correlation dynamics in European equity markets. Discussion Paper, The Institute for International Integration Studies.

Khandani, A., and A. Lo. 2007. What happened to the quants in August 2007? Working Paper, available at http://ssrn.com/abstract=1015987 (November 4, 2007).

Kim, S.-J., F. Moshirian, and E. Wu. 2006. Evolution of international stock and bond market integration: influence of the European monetary union. *Journal of Banking & Finance* 30:1507-34.

Kolev, N., U. dos Anjos, and B. V. de M. Mendes. 2006. Copulas: a review and recent developments. *Stochastic Models* 22:617-60.

Koutmos, G. 1999. Asymmetric index stock returns: evidence from the G-7. *Applied Economics Letters* 6:817-20.

Kroner, K. F., and V. Ng. 1998. Modeling asymmetric comovements of asset returns. *Review of Financial Studies* 11:817-45.

Ku, Y.-H. H., H.-C. Chen, and K.-H. Chen. 2007. On the application of the dynamic conditional correlation model in estimating optimal time-varying hedge ratios. *Applied Economics Letters* 14:503-9.

Kuper, G. H., and Lestano. 2007. Dynamic conditional correlation analysis of financial market interdependence: an application to Thailand and Indonesia. *Journal of Asian Economics* 18:670-84.

Lando, D. 2004. *Credit Risk Modeling*. Princeton University Press.

Ledford, A. W., and J. A. Tawn. 1996. Statistics for near independence in multivariate extreme values. *Biometrika* 83:169–187.

——. 1997. Modelling dependence within joint tail regions. *Journal of the Royal Statistical Society* B 59:475–99.

——. 1998. Concomitant tail behaviour for extremes. *Advances in Applied Probability* 30:197–215.

Ledoit, O., and M. Wolf. 2003. Improved estimation of the covariance matrix of stock returns with an application to portfolio selection. *Journal of Empirical Finance* 10:603–621.

Ledoit, O., P. Santa-Clara, and M. Wolf. 2003. Flexible multivariate GARCH modeling with an application to international stock markets. *Review of Economics & Statistics* 85:735–47.

Lee, J. 2006. The comovement between output and prices: evidence from a dynamic conditional correlation GARCH model. *Economics Letters* 91:110–16.

Lehmann, B. 1990. Fads, martingales, and market efficiency. *Quarterly Journal of Economics* 105:1–28.

Lien, D., Y. K. Tse, and A. K. C. Tsui. 2002. Evaluating the hedging performance of the constant-correlation GARCH model. *Applied Financial Economics* 12: 791–98.

Lo, A., and A. C. MacKinlay. 1990a. An econometric analysis of nonsynchronous trading. *Journal of Econometrics* 40:203–38.

——. 1990b. When are contrarian profits due to stock market over-reaction? *Review of Financial Studies* 3:175–206.

Longin, F., and B. Solnik. 2001. Extreme correlation of international equity markets. *Journal of Finance* 56:649–76.

Longstaff, F., and A. Rajan. 2008. An empirical analysis of the pricing of collateralized debt obligations. *Journal of Finance* 63:529–64.

Maghrebi, N., M. J. Holmes, and E. J. Pentecost. 2006. Are there asymmetries in the relationship between exchange rate fluctuations and stock market volatility in Pacific Basin countries? *Review of Pacific Basin Financial Markets & Policies* 9:229–56.

McAleer, M., and B. da Veiga. 2008. Forecasting Value-at-Risk with a parsimonious portfolio spillover GARCH (PS-GARCH) model. *Journal of Forecasting* 27:1–19.

McNeil, A., R. Frey, and P. Embrechts. 2005. *Quantitative Risk Management.* Princeton University Press.

Merton, R. C. 1974. On the pricing of corporate debt: the risk structure of interest rates. *Journal of Finance* 29:449–70.

Michayluk, D., P. J. Wilson, and R. Zurbruegg. 2006. Asymmetric volatility, correlation and returns dynamics between the U.S. and U.K. securitized real estate markets. *Real Estate Economics* 34:109–31.

Milunovich, G., and S. Thorp. 2006. Valuing volatility spillovers. *Global Finance Journal* 17:1–22.

Nelson, D. B. 1991. Conditional heteroskedasticity in asset returns—a new approach. *Econometrica* 59:347–70.

Nelson, D. B., and D. P. Foster. 1994. Asymptotic filtering theory for univariate ARCH models. *Econometrica* 62:1–41.

Newey, W. K., and D. McFadden. 1994. Large sample estimation and hypothesis testing. In *Handbook of Econometrics* (ed. R. F. Engle and D. McFadden), volume 4, pp. 2113–245. Elsevier.

Ng, V., R. F. Engle, and M. Rothschild. 1992. A multi-dynamic-factor model for stock returns. *Journal of Econometrics* 52:245–66.

Niguez, T.-M., and A. Rubia. 2006. Forecasting the conditional covariance matrix of a portfolio under long-run temporal dependence. *Journal of Forecasting* 25: 439–58.

Palandri, A. 2005. Sequential conditional correlations: inference and evaluation. Working paper, Social Science Research Network.

Patton, A. J. 2006a. Estimation of multivariate models for time series of possibly different lengths. *Journal of Applied Econometrics* 21:147–73.

——. 2006b. Modelling asymmetric exchange rate dependence. *International Economic Review* 47:527–56.

Pelagatti, M. 2004. Dynamic conditional correlation with eliptical distributions. Available at SSRN: http://ssrn.com/abstract=888732 (August 24, 2004).

Pelletier, D. 2006. Regime switching for dynamic correlations. *Journal of Econometrics* 131:445–73.

Poon, S.-H., M. Rockinger, and J. Tawn. 2004. Extreme value dependence in financial markets: diagnostics, models, and financial implications. *Review of Financial Studies* 17:581–610.

Rabemananjara, R., and J. M. Zakoian. 1993. Threshold ARCH models and asymmetries in volatility. *Journal of Applied Econometrics* 8:31–49.

Rangel, J. G., and R. F. Engle. 2008. The factor Spline GARCH model. Manuscript, Department of Finance, New York University.

Resnick, S. 2004. The extremal dependence measure and asymptotic independence. *Stochastic Models* 20:205–27.

Rosenblatt, M. 1952. Remarks on a multivariate transformation. *Annals of Mathematical Statistics* 23:470–72.

Rosenow, B. 2008. Determining the optimal dimensionality of multivariate volatility models with tools from random matrix theory. *Journal of Economic Dynamics & Control* 32:279–302.

Ross, S. 1976. The arbitrage theory of capital asset pricing. *Journal of Economic Theory* 13:341–60.

Samuelson, P. A. 1965. Proof that properly anticipated prices fluctuate randomly. *Industrial Management Review* 6:21–49.

Scholes, M., and J. Williams. 1977. Estimating beta from non-synchronous data. *Journal of Financial Economics* 5:309–27.

Schönbucher, P. 2003. *Credit Derivatives Pricing Models.* New York: Wiley.

Sharpe, W. 1964. Capital asset prices: a theory of market equilibrium under conditions of risk. *Journal of Finance* 19:425–42.

Silvennoinen, A., and T. Terasvirta. 2005. Multivariate autoregressive conditional heteroskedasticity with smooth transitions in conditional correlations. Working Paper Series in Economics and Finance, Stockholm School of Economics.

——. 2008. Multivariate GARCH models. In *Handbook of Financial Time Series* (ed. T. W. Andersen, R. A. Davis, J. P. Kreiss, and T. Mikosch). Springer.

Skintzi, V. D., and S. Xanthopoulos-Sisinis. 2007. Evaluation of correlation forecasting models for risk management. *Journal of Forecasting* 26:497–526.

Sklar, A. 1959. Fonctions de répartition à *n* dimensions et leurs marges. *Publications de l'Institute de Statistique de l'Université de Paris* 8:229–31.

Storti, G. 2008. Modelling asymmetric volatility dynamics by multivariate BL-GARCH models. *Statistical Methods & Applications* 17:251–74.

Styan, G. P. H. 1973. Hadamard products and multivariate statistical analysis. *Linear Algebra and Its Applications* 6:217–40.

Thorp, S., and G. Milunovich. 2007. Symmetric versus asymmetric conditional covariance forecasts: does it pay to switch? *Journal of Financial Research* 30: 355–77.

Tse, Y. K., and A. K. C. Tsui. 2002. A multivariate generalized autoregressive conditional heteroscedasticity model with time-varying correlations. *Journal of Business & Economic Statistics* 20:351–62.

UBS. 2008. Shareholder report on UBS's write-downs.

Van der Weide, R. 2002. GO-GARCH: a multivariate generalized orthogonal GARCH model. *Journal of Applied Econometrics* 17:549–64.

White, H. 1984. *Asymptotic Theory for Econometricians*. Academic Press.

Wooldridge, J. M. 1994. Estimation and inference for dependent processes. In *Handbook of Econometrics* (ed. R. F. Engle and D. McFadden), volume 4. Elsevier.

Yang, L., F. Tapon, and Y. Sun. 2006. International correlations across stock markets and industries: trends and patterns 1988–2002. *Applied Financial Economics* 16:1171–83.

Yu, J., and R. Meyer. 2006. Multivariate stochastic volatility models: Bayesian estimation and model comparison. *Econometric Reviews* 25:361–84.

Index